The Role of Markets in the World Food Economy

Also of Interest

†Available in hardcover and paperback.
*Available only in paperback.

Westview Special Studies in Agriculture Science and Policy

The Role of Markets in the World Food Economy
edited by D. Gale Johnson
and G. Edward Schuh

This book extends the discussion of world food problems by giving explicit recognition to the potential role of markets. The authors highlight the contribution of prices to the solution of food problems in low-income countries, for example, by providing adequate incentives to farmers to expand production, assuring that food supplies can be obtained through trade when needed and giving appropriate signals to consumers. They also document the negative effects on food supply and national welfare of the actual price policies of many Third World governments.

While recognizing the problems involved in defining and measuring hunger, as well as in improving the food supply, the authors consider the outlook for future food availability as favorable in terms of continued modest improvement in per capita food supplies at prices, adjusted for inflation, that are likely to continue the slow decline of recent decades. One focus of their comments is the positive roles that governments can and should play in the world food economy, especially in support of research, creation of human capital, and provision of appropriate rural infrastructure.

D. Gale Johnson is professor and chairman of the Department of Economics at The University of Chicago. **G. Edward Schuh** is professor and chairman of the Department of Agriculture and Applied Economics at the University of Minnesota, St. Paul.

The Role of Markets in the World Food Economy

edited by D. Gale Johnson
and G. Edward Schuh

Routledge
Taylor & Francis Group

LONDON AND NEW YORK

First published 1983 by Westview Press

Published 2019 by Routledge
52 Vanderbilt Avenue, New York, NY 10017
2 Park Square, Milton Park, Abingdon, Oxon OX14 4RN

Routledge is an imprint of the Taylor & Francis Group, an informa business

Library of Congress Cataloging in Publication Data
Main entry under title:
The role of markets in the world food economy.

    (Westview special studies in agriculture science and policy)
    Includes index.
    1. Food supply—Government policy—Addresses, essays, lectures. 2. Agriculture
and state—Addresses, essays, lectures. 3. Food prices—Government policy—Addresses,
essays, lectures. I. Johnson, D. Gale (David Gale), 1916– . II. Schuh, George Edward,
1930– . III. Series.
HD9000.6.R73   1983   338.1'9   83-6518

ISBN 13: 978-0-367-29564-6 (hbk)
ISBN 13: 978-0-367-31110-0 (pbk)

Contents

Preface

The papers and discussions in this book are revised versions of the presentations made at a conference held in Minneapolis on October 15–16, 1982. The conference, which bore the same title as this volume, namely, *The Role of Markets in the World Food Economy,* was jointly sponsored by the Department of Economics of The University of Chicago and the Department of Agricultural and Applied Economics of The University of Minnesota.

The invitation that was sent to a selected list of individuals and organizations by the coeditors of this volume described our objectives in this way:

> This conference—made possible by a grant from Cargill, Incorporated—is designed to explore the role markets do and can play in the world food economy. The issue is both important and complex. It has attracted interest and study from many quarters—academics, public officials and governments, international public institutions, church and social action groups, farmers, agribusiness, media and philanthropic organizations.
>
> Typically, however, these different groups have little opportunity to discuss these issues with each other, to learn from each other and to develop a broader consensus on and understanding of the world food economy. The purpose of this conference is to provide that opportunity. Our goal is not to agree on recommendations but to have a dialogue among responsible people from these different backgrounds that might shape future individual involvement, public thought and action.

It is our hope that this volume will serve as the basis for a continuing dialogue among those who have concerns about the future well-being of the half of the world's population that lives in the low-income developing countries of the world. It is clear from the papers and discussions that the world has the resources to improve the lot, with respect to all of the material aspects of their lives, of the more than 2 billion people living in the low-income countries. It is, we hope, also clear that the extent of the improvement that will occur in the rest of this century will depend upon the policies followed by most governments, including the governments of the industrial countries. Increasing the reliance upon effectively functioning markets will make an important contribution to improving the lot of the poor people of the world. But no claim is made in any of the papers that any one measure or particular change in policy would be sufficient to result in a major improvement. However, it does seem to be the case that one or two very adverse policies, such as exceedingly low prices for important farm products, can nullify the effects of other quite positive policies.

We want to express our appreciation to those who contributed papers and discussions to the conference. As can be seen from the contents of this book, the authors were serious and responsible in their work. Whatever merit this volume may have is due to the quality of the individual contributions.

D. Gale Johnson

Part 1 _____

1

The World Food Situation: Recent and Prospective Developments

D. Gale Johnson
Department of Economics
The University of Chicago

My assignment is to review the major trends in world food production, trade, and prices for the past two decades. During those two decades there were two periods when fears were expressed that world food supplies were insufficient to prevent wide-scale suffering and loss of life. One such time was in the early and mid-1960s; the second came less than a decade afterward. This paper explores the major reasons why production, trade, and price developments after 1972 did *not* result in a deterioration in the world food situation but actually provided some modest improvement for the world's poor people. I shall conclude by considering some of the factors that may influence prices and the availability of food during the 1980s.

Some Recent Views

Anyone who is at all observant knows that there is significant disagreement among the numerous reports about the current state of the world food system. Estimates of the number of seriously malnourished persons in the world range from a few tens of millions to a billion or more (see Poleman, Chap. 2 this book). With such uncertainty existing, it is not surprising that there are equal or greater differences in views concerning expected developments for the next decade or the remainder of the century. It is not my intent to review in detail the projections of trends in the demand and supply of food. I shall, however, present a range of views to provide the flavor of the existing state of opinion on these matters.

But I have no intention of concealing what I believe, namely, that the weight of the evidence supports the following points:

1. There has been modest improvement in food supplies per capita in the low-income countries of the world during the past three decades.

2. The real prices of the major sources of calories for poor people—the grains—have declined in recent decades rather than increased, and at the present time the real international prices for grains, vegetable oils, and sugar are at or near all-time low levels.

3. International trade in grain has grown rapidly during the past decade, but the low-income countries have not contributed significantly to the import growth as had been feared they would at the time of the World Food Conference.

4. It is reasonable to expect that the international prices of grains and vegetable oils will remain at historically low and perhaps declining levels for the next one or two decades.

5. Resources are available that would permit increasing the rate of growth of per capita food production in the developing countries.

6. A world food system has evolved over the past three decades that has substantially reduced the risks of food shortages and famine and hunger resulting from natural disasters, and by the end of this century nearly every person in the world could have access to the world food system if that person had adequate income to purchase the food that is available; at present a much larger percentage of the world's people could be participants in the world food system than is now the case, if governments did not intervene in the movement of food.

I shall go on now to the various projections being made with regard to the broad trends in world food supply, demand, prices and international trade.

The "Things Are Getting Worse" Scenario

The Global 2000 Report to the President, a major study undertaken by the U.S. government at the request of President Carter, painted a dark picture for the future. It concluded (Barney 1982, p. 17): "After decades of generally falling prices, the real price of food is projected to increase 95 percent over the 1970–2000 period, in significant part as a result of increased energy dependence." An increase in food prices at an annual rate of 2.25 percent for three decades is, indeed, a startling projection. Nothing approaching such a rate of increase for an extended period has occurred in the past several centuries. In the text of the report three different scenarios are presented, each with a projection of a range of prices or a particular price for the year 2000. The lowest of the projections is for an increase of 35 percent, and the highest is for an increase of 115 percent, while the base projection indicates prices may increase by 45 to 95 percent (p. 96). I found absolutely nothing in the text that indicated that any one of the price projections was considered more likely than any of the others. It is not clear why the price projections near the high end of the alternatives were chosen.

Global 2000 makes another startling projection—a sharp break with the trends of the twentieth century: "A closer look at the projections suggests that a substantial increase in the share of the world's resources committed to food production will be needed to meet population- and income-generated growth in demand through 2000" (Barney 1982, p. 96). If this statement has any meaning, it is that between now and 2000 the share of the world's income devoted to the purchase of food will increase. Such has not been the case for many decades; it is not obvious why past trends toward a diminution of the share of the world's income devoted to food should be reversed. During the 1970s for the world as a whole and for every major grouping of countries the share of agriculture in total employment declined. For the world the employment share of agriculture declined from 51.3 percent in 1970 to 45.3 percent in 1980 (FAO 1980). Given that the real price of food at wholesale was less in 1980, admittedly rather slightly, than in 1970, the projected increase in the share of the world's resources devoted to food must all come in the two decades between 1980 and 2000. The projected increase in the world market prices of food of 95 percent must also come in two decades, implying the remarkable annual growth rate of food prices of 3.4 percent! It is difficult to understand why price developments that occurred during the 1970s could not have been reflected, at least in part, in a study that was published in 1980.

Patrick O'Brien of the U.S. Department of Agriculture (USDA) is the primary custodian of the USDA grain-oilseed-livestock (GOL) model that was used in *Global 2000* for the price projections for 1985 and 2000. In an article published in 1981 he summarized the results of his analysis as follows:

Should this prognosis prove correct, the early eighties will bring two fundamental changes in the U.S. agriculture. First, annual increases in foreign and domestic demand due to population and income growth will be greater, on average, than the increases in productive capacity due to resource expansion and productivity gains. The real prices received by farmers—given normal weather—should increase; scenarios generated using several longrun equilibrium simulation models suggest real price increases of 1 to 3 percent per year, compared with declines averaging 1 to 2 percent per year for the postwar period to date. Moreover, if the gains in capacity in the early eighties due to productivity and resource growth are more than offset by the losses in capacity resulting from unit cost increases and increasingly stringent environmental constraints, the real prices associated with the output needed to balance foreign and domestic demand could be substantially higher (p. 20).

The time period to which the annual real price increases of 1 to 3 percent are to apply is the first half of the 1980s. The tightening of the demand-supply relationship referred to clearly had not occurred by mid-1982.

John Mellor, director of the International Food Policy Research Institute (IFPRI), has accepted the view that real food prices are likely to increase during the 1980s (IFPRI 1982, p. 7). In November 1981 he stated:

In commenting on the global food situation, I want to present a somewhat different picture from the standard one of the last few years. Third World countries will increase their food imports in the next few decades by so much that they themselves will bring about tight global supplies and rising real food prices.

This situation is the result of rapidly rising real incomes. I think it is important to keep in mind that this income growth is what will dictate the food problem in the next few decades. Currently, some 700 million people in Third World countries are experiencing an extraordinary 4 percent or more growth rate of per capita income. That is more rapid growth than Europe was able to sustain in the post–World War II recovery period. The portion of the population sharing in this rapid growth could easily double in the next few years.

Demand for food in these countries is growing at 4–6 percent a year. I am talking about effective demand—what people are able and willing to pay—and about rates of growth of demand that are rarely exceeded, even in countries that have good agricultural research systems and are following optimal agricultural development policies.

It is reasonable to conclude that the rapidly growing middle-income developing countries will increase their grain and food imports in the years ahead. The source of the growth of demand for imports is now said to be the middle-income developing countries and not, as was the case less than a decade ago, low-income countries such as India and Pakistan. But as I will show later, some of the sources of rapid growth of food imports during the 1970s are likely to show much slower growth during the 1980s.

The "Things Are Improving Slowly" Scenario

I classify the major report of the Food and Agriculture Organization, *Agriculture: Toward 2000*, as falling in the "Things Are Improving Slowly" scenario. This is my classification, not FAO's, and FAO might mildly object. But the report indicates that if recent trends are continued, per capita food supplies in the 90 developing countries would increase by almost 9 percent between 1980 and 2000 and that increases of 15 to 20 percent in per capita food consumption could be achieved. And while gross cereal imports of the developing countries are projected to increase from 86 million metric tons in 1980 to from 146 to 220 million metric tons, trend agricultural production would be at an annual rate of 2.8 percent compared to a trend projection of demand of 2.9 percent. The difference between 2.8 percent and 2.9 percent is so small that the increase in cereal imports would more than provide for the production shortfall. Net cereal imports have trend values for 2000 of 165 million m. tons (exports of 61 million m. tons).

The trend analysis provides this striking conclusion (FAO 1981, pp. 24–25):

A continuation of production and demand trends world-wide implies large global surpluses in some commodities, despite the growing millions of seriously undernourished, and deficits in other, mainly livestock products. Thus the net surplus of cereals of the developed countries would tend to stand at 213 million tonnes at a time when the developing countries would have a deficit of 165 million tonnes. Projected net availabilities from the developing countries of such competing products as sugar, citrus fruit and vegetable oils and oilseeds would substantially outstrip import demand in the developed countries, whose continued protectionism would limit severely any expansion in their imports of these products.

All in all, the tendency is for global surpluses worth roughly $20 billion$_{1975}$. The developed countries as a group would tend to become a substantial net agricultural exporter; growing exports from their cereals and livestock sectors would not be offset by similar expansions in imports of other commodities, either competing or tropical.

At the same time, the trend is for the developing countries to have production balances sufficient to maintain their position as net agricultural exporters although their overall agricultural trade surplus would continue to decline. But this outcome implies that their net exports to the developed countries of competing and noncompeting tropical products increase sufficiently to offset increases in the imports of cereals and livestock products, and this is contrary to the projected import trends of developed countries. Obviously, therefore, the trend projections of the developed and developing countries are incompatible.

The projected imbalances will not actually materialize; spontaneous or policy-induced adjustments will bring balance. The policy issue is how orderly adjustments can be brought about. They should increase rather than reduce trade—the developing countries should be able to import and consume more of the cereals produced in the developed countries and help to pay for them with increased exports of the commodities they are best suited to produce. Such an evolution would eliminate the irony in the situation of a world afflicted with under-nutrition, but where there is a danger of global surpluses.

The seriousness of the deterioration in the production and import trends of developing countries comes out most strongly in the self-sufficiency ratios (SSRs) for cereals, their major agricultural import. For every region and every income category of developing countries this deterioration holds. In Latin America and the Far East, the decline is only 3 to 4 percent, but with an SSR for 1980 put equal to 100, the year 2000 index is 87 for the Near East and only 74 for Africa. For low-income countries, the index falls to 93, while for the least developed group of countries it is barely 80.

But the most optimistic aspect of the FAO report consists of the outlines of feasible alternatives to the trend projections. Either of the major alternatives, if realized, would permit significant further improvement in income growth, agricultural output, per capita food supplies, a reduction in cereal imports and an increase in the net surplus in agricultural trade for the 90 developing countries. The two scenarios that were developed indicate that annual agricultural output growth rates for the low-income developing countries could be increased from the 2.7 percent to 3.1 to 3.8 percent.[1] For the 41 low-income developing countries the scenario that could result in a 3.1 percent annual growth of agricultural production would provide an increase in calories sufficient to increase per capita consumption from 89 percent of average requirements for 1974–1976 to 103 percent in 2000; if agricultural output grew at the trend rate of 2.7 percent there would have been significant nutritional improvement and caloric intake would reach 97 percent of average requirement.

I do not want to imply that *Agriculture: Toward 2000* provides a basis for complacency; it was certainly not so intended by its authors. But the report can, and I believe should, be interpreted as indicating that the resources—physical and human—exist to provide for a significant improvement in the nutritional status for the vast majority of the population of the developing countries. Even with the trend projections of agricultural production and incomes the percentage of the population of the developing countries (excluding China) classified as "seriously undernourished" would decline from 23 percent in 1975 to 17 percent in 2000. However, with the scenario providing for modest improvement over trend the percentage malnourished in 2000 would decline to 11 percent; with the more optimistic scenario the decline would be to 7 percent.

I want to emphasize that those of us who are cautiously optimistic about the continued improvement in the consumption levels of most of the poor people of the world do not believe there is any room for complacency. If circumstances are to improve, it will be because efforts are made to make the improvement occur, and because at least some of the hindrances that exist—trade restrictions, low farm prices due to governmental constraints, and inadequate provision of farm inputs— are ameliorated. Thus to say that things are going to get better does not mean that this will be the outcome if nothing is done; what it means is that with the resources devoted to food production and with the policies that are likely to be followed, improvement is likely to occur. Under more favorable circumstances, the rate of improvement could be increased. The two alternative scenarios presented in *Agriculture: Toward 2000* indicate what is possible with added effort and modest policy changes.

The Recent Past

During the 1960s and 1970s the developing countries as a group had a modest improvement in per capita food supplies. However, according to FAO estimates the increase from 1961–1965 to 1974–1976 was only 3 percent for the period and was unevenly distributed among the developing countries (FAO 1981). The 41 countries with per capita gross domestic product (GDP) of less than $300 (1975) had constant per capita supplies of calories while the middle-income developing countries (all others) had an increase of just 7 percent.

For the developing world, per capita food production increased by 14 percent between 1951–1953 and 1979–1981 (USDA 1981). This represents a quite modest growth rate of 0.46 percent. However, the growth rate for total food production looks very impressive at 2.89 percent annually. This compares to an output growth rate of food production in the developed market economies of 2.1 percent. Given the subsidies provided for the agricultures of the developed countries,

TABLE 1

Life Expectancy at Birth, 1950, 1960 and 1978

	1950	1960	1978	Increase, 1950-78
	------------(years)------------			
Industrialized countries	66.0	69.4	73.5	7.5
Middle-income countries	51.9	54.0	61.0	9.1
Low-income countries	35.2	41.9	49.9	14.7
Centrally planned economies[a]	62.3	67.1	69.9	7.6

Source: World Bank, World Development Report, 1980, p. 34.

[a]Excludes China.

the agricultural performance of the developing countries looks good indeed.

The modest growth in food production per capita for the developing world reflects the very poor performance of Africa for three decades and the very slow growth in southern Asia during the 1970s. Africa had a constant average level of per capita food production during the 1950s and 1960s and a shocking decline during the 1970s. In 1980 per capita food production in Africa (excluding South Africa) was 15 percent below 1969-1971 levels. Total food production increased 10 percent while population grew by about 25 percent, resulting in an unprecedented decline in per capita food production.

The only slightly better recent performance of southern Asia compared to Africa may be a greater source of concern. There are many more people in southern Asia than in Africa, and there are relatively limited potentials for further land developments there, while Africa has significant potentials for land development. Consequently most of the increased output of food in southern Asia must come from higher yields per unit of land. This is not entirely unfortunate, as the experience of the last three decades in the industrial countries as well as in some of the developing countries has indicated that expanding output through higher yields costs less than bringing new land into cultivation.

Perhaps the best indicator of the state of nutrition and health has been the continuing increase in life expectancy among all of the developing regions and at all national income levels. In fact, the data on life expectancy cast some doubts that the improvements in per capita food supplies and nutrition were as modest as the aggregate food data indicate, and may cast even greater doubts on the estimated declines in and low level of per capita food consumption in many African countries. Table 1 compares changes in life expectancies for groups of countries for the years from 1950 to 1978. Low-income countries had less than $360 per capita gross national product (GNP) in 1978 and had a population of 1.29 billion; the middle-income developing countries had a population of 873 million.

The largest absolute increase in life expectancy at birth between 1950 and 1978 occurred in the low-income countries. The increase was 14.7 years; in 1978 the life expectancy in these countries was nearly 50 years—about the same level as that of the middle-income developing countries in 1950.

The World Bank has presented estimates of life expectancy for African countries as of mid-1979 for different regions and country income levels. These data, presented in Table 2, are compared with developing countries generally and with specific developing countries. Life expectancy in sub-Saharan Africa is somewhat lower than for all low-income developing countries (excluding centrally planned economies). But the difference is not very large—47 years for all of sub-Saharan Africa and 50 for all low-income developing countries. Higher per capita incomes in Africa do not, as yet, seem to have brought as large an increase in life expectancy as the same income differentials have in other regions.

Post-1972 Developments Affecting World Food

The information brought together in the previous section, as well as the more detailed data and analysis included in Thomas Poleman's paper, indicate that for most of the low-income countries of the world, food supplies and nutrition improved during the 1970s. Such improvement was contrary to the widely expressed views that developments in food demand, supply, costs, and prices opened a new era of scarcity and human suffering, with the latter on an unparalleled scale. There was talk of the necessity of triage with respect to access to the world's food supply and of the overcrowded lifeboat analogy, as though the world's capacity to produce food were fixed.

The summary of an article published in *Science* (12 December 1975) stated quite accurately what was the "conventional wisdom" of the day as perceived by most of the media organizations:

> The scarcity of basic resources required to expand food output, the negative ecological trends that are gaining momentum year by year in the poor countries, and the diminishing returns on the use of energy and fertilizer in agriculture in the industrial countries lead me to conclude that a world

TABLE 2

Life Expectancy at Birth, African and Other Developing Countries, 1960 and 1979

Country Group	Life Expectancy (Years)			1979 GNP Per Capita (U.S. Dollars)	Index of Per Capita Food Production 1977-79 (1969-71 = 100)	Daily Calorie Supply Per Capita, 1977	
	1960	1979	Increase			Calories	Percent of Requirement
Africa[a]							
Low Income	38	46	8	239	91	2,040	91
Low income, semi-arid	37	43	6	187	88	1,992	89
Low income, other	39	47	8	247	91	2,086	93
Middle income oil importers	41	50	9	532	95	2,180	97
Middle income oil exporters	39	48	9	669	86	1,970	89
Sub-Saharan Africa	--	47	-	411	91	--	--
Selected Low Income Countries	42	50	8	200	97	2,052	91
India	43	51	8	180	100	2,021	91
Bangladesh	40	47	7	90	90	1,812	78
Developing Countries by Per Capita Income							
Less than $390	42	50	8				
$390 - 1,050	46	55	9				
$1,060 - 2,000	47	64	7				
$2,040 - 3,500	65	71	6				

Source: World Bank, World Development Report 1981, Washington, D.C., 1981, pp. 134, 168, and Accelerated Development in Sub-Saharan Africa: An Agenda for Action, Washington, D.C.: World Bank, 1981, pp. 143, 177.

[a]Excludes South Africa.

of cheap, abundant food with surplus stocks and a large reserve of idled cropland may now be history. In the future, scarcity may be more or less persistent, relieved only by sporadic surpluses of a local and short-lived nature. The prospects are that dependence on North America will be likely to continue to increase, the increase probably being limited only by the region's export capacity.

What happened during the last half of the 1970s and the early 1980s that negated, at least up to now, the gloomy prediction just presented? It is to this matter that I now turn. I shall first consider price trends of the 1970s and for earlier periods, in some cases for the years since 1910.

Price Trends

In 1972 and 1973 there were numerous predictions or projections that the world was faced with a sharp reversal of the declining trend in the prices of the major food products. Lester Brown (1981), among others, believed that a new price era had emerged: "Food prices are likely to remain considerably higher than they were during the past decade." Such a conclusion as this could be described as the "conventional wisdom"; dissenters from this view were little noticed by either the media or the policymakers.

Figures 1 through 4 present real or deflated prices for important food commodities. Figures 1 and 2 give the real prices for wheat and corn, based on U.S. export unit values, starting with 1910 for selected periods and annually since 1960. Figures 3 and 4 give the real prices of two other important sources of calories—soy oil and sugar—annually for 1950 to date. The price of soy oil reflects rather closely the prices of all of the important vegetable oils that are used for food. Figure 5 includes the deflated U.S. export unit values of cotton. In many parts of the world the production of cotton is competitive with the production of cereal crops; in addition, of course, the price and cost of cotton is important as a component of living costs for low-income families in most areas of the world.

The conversion from current prices in U.S. dollars to deflated or real dollars has been made through the use of the U.S. wholesale price index. The World Bank publishes data on the prices of farm and mineral products as well as of farm inputs such as fertilizer. The deflator used by the World Bank is the CIF measure (cost, insurance, and freight) of unit values of imports of manufactured products by the developing countries. The index used by the World Bank increases at a substantially greater rate over time than does the U.S. wholesale price index. Compared to 1967 the World Bank index was 243 in 1977, while the number was

FIGURE 1

U.S. WHEAT EXPORT PRICES
SELECTED YEARS 1910–1981
DEFLATED BY 1967=100 W.P.I.

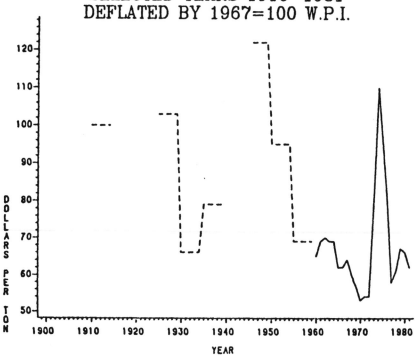

The data presented in Figs. 1-5 do not reflect the income
payments made under the various farm programs in the U.S.
Relative to prices received by farmers, the payments were
approximately three times higher from 1967 to 1971 than
from 1977 to 1981 for corn, wheat, and cotton. There
were no payments for soybeans or soy oil. The sugar
price is an international price and does not reflect
prices received by U.S. farmers. The export prices are
deflated by the U.S. wholesale price index with a base of
1967 equal to 100. The export prices and the wholesale
price index (W.P.I.) are from the U.S. Dept. of Commerce,
Bureau of the Census, Statistical Abstract of the United
States, Washington, D.C., annual issues.

FIGURE 2

U.S. CORN EXPORT PRICES
SELECTED YEARS 1910–1981

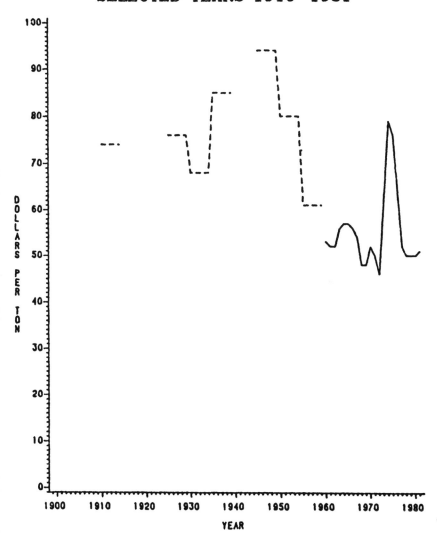

DEFLATED BY 1967=100 W.P.I.

See footnote to Figure 1.

FIGURE 3

DECATUR SOY OIL PRICES
1950–1981

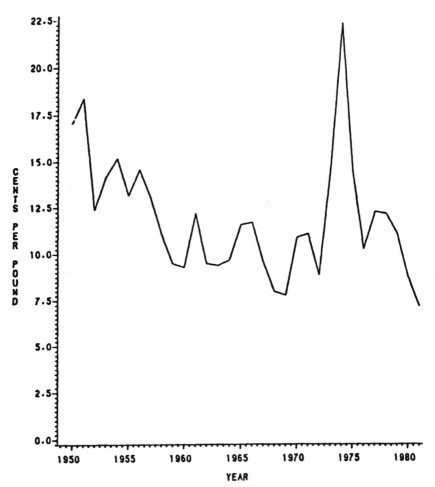

DEFLATED BY 1967=100 W.P.I.

See footnote to Figure 1.

FIGURE 4

RAW SUGAR WHOLESALE PRICES
CARRIBEAN PORTS
1950–1981

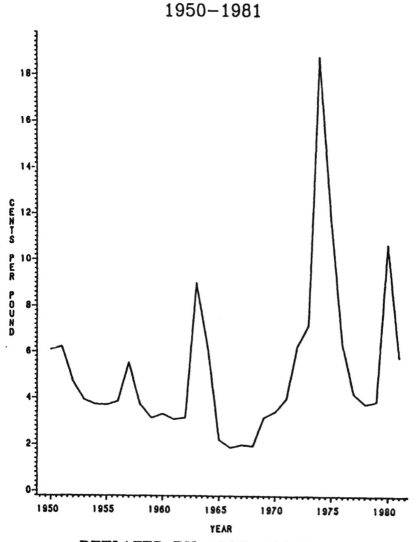

DEFLATED BY 1967=100 W.P.I.

See footnote to Figure 1.

FIGURE 5

U.S. COTTON EXPORT PRICES
1950-1981

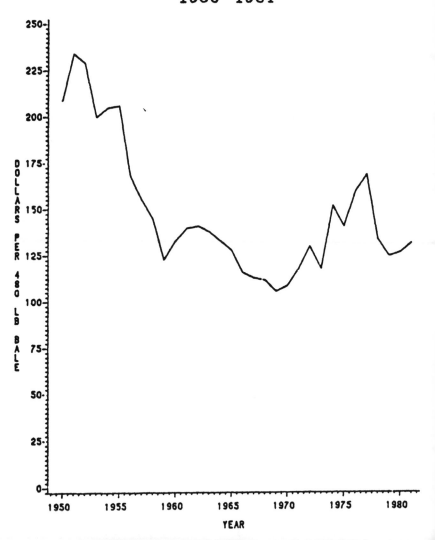

DEFLATED BY 1967=100 W.P.I.

See footnote to Figure 1.

193 for the wholesale price index. The use of the World Bank index would have reduced real prices in 1977 by 20 percent, compared to the use of the U.S. wholesale price index.

The price trends we have presented are for commodities that are traded internationally in significant amounts. The real prices were generally lower or the same at the end of the 1970s and the beginning of the 1980s than at the beginning of the 1970s. One possible explanation of such trends could be that international trade in these products lagged or was actually less at the end of the decade than at the beginning, but this explanation has no validity—trade in grain nearly doubled, as did the trade in soybeans. In spite of the trade restrictions that limit international trade in sugar, trade increased by 25 percent during the decade. World trade in vegetable oils more than doubled, increasing by 148 percent from 1969–1970 to 1979–1980 (from 4.4 million to 10.9 million m. tons).

The real prices of four of the five agricultural products reached peak levels for the 1960s and 1970s in 1974; cotton was the exception, reaching its peak real price in 1977. But it is worthy of note that even the high prices of 1974—up by 70 percent or more from the levels of the early 1970s—were of the same order for wheat and corn as the real prices of the late 1920s and below the 1945–1949 prices.

For the four commodities that reached recent peak prices in 1974, price declines occurred suddenly and dramatically. The real price of wheat fell by almost half in three years, corn by more than one-third in four years, sugar by 80 percent in four years, and soy oil by more than half in two years. The prices of the mid-1970s that were thought to represent a new and higher level very soon declined to the levels of the early 1970s. Thus the price experience of the late 1970s and the early 1980s has not borne out the expectation that the real costs of food products would increase significantly after the mid-1970s.

Cotton prices exhibited greater price stability during the 1970s than did any of the four food products. The peak of real prices was not reached until 1977, the year that was the low point for wheat and nearly so for the other three products. The 1977 peak was 55 percent over the 1970 low.

It should come as no surprise that cotton prices were the most stable and sugar prices the least stable. Of the five commodities, cotton was least subject to protection and governmental intervention while sugar was the most protected and interfered with. According to estimates made by Valdés and Zietz (1980), wheat was the most protected of the remaining three and corn the least. Wheat prices were significantly more unstable than corn prices, though less unstable than soy oil. With the exception of soy oil, the instability of the prices of the other four products was consistent with expectations derived from the degree of protection and governmental intervention in the markets.

In considering the role of the market in the world food economy, it is appropriate to note that one of the important sources of price instability is governmental intervention. In particular, when governments stabilize domestic prices of food products, they contribute to international price instability. Protecting producers and consumers from changes in world supply and demand conditions means that all of the adjustment must be made in that part of the world market where consumers and producers react to prevailing world market prices, or to prices that vary to some degree with the world prices (Johnson 1975a).

In the remainder of my paper I shall review the arguments that were made during the early 1970s in support of the projection that real grain and other food prices would be at a new and higher level after the events of 1973 and 1974. Because similar arguments are being made once again, the discussion is not primarily a matter of flogging a dead horse. It remains important to determine if there are significant trends under way that could cause real food prices to increase in the next decade or two. It is always possible that the adverse effects of a particular factor have not yet occurred but may do so in the near future. In addition, two rather new concerns—one that might have increased the demand for land and other agricultural resources and one that decreases the supply of land—have attracted significant attention in recent years.

Feed Versus Food

During 1973 and 1974 it was commonplace to view with alarm the increased use of grains for livestock feed. It was concluded by some that increasing incomes—affluence—in both the industrial countries and in the rapidly growing middle-income developing countries would increase the feed use of scarce grains at such a rapid rate as to reduce the amount of grain available for low-income populations.

In a monograph written in early 1975, I argued that the reverse was actually the case—that increasing affluence in the industrial countries increased the supply of grain available to low-income countries because affluence increased supply as well as demand (Johnson 1975b). Since World War II it was quite clear that rising productivity—the primary source of increasing affluence—had added more to the supply of grains than to demand in the industrial countries. The proof for this statement was the sharp increase in net exports by the industrial market economies. In 1960–1962 the grain exports from the industrial market economies were but 20.3 million metric tons; in 1969–1971 exports had increased to 31.9 million m. tons and could have been substantially greater had the demand been greater. During those years the major exporters were actually engaged in reducing grain production.

In response to the sharp increase in world grain prices, grain exports from the industrial market economies increased to nearly 60 million metric tons in both 1972/1973 and 1973/1974. By the end of the 1970s and the beginning of the 1980s, net grain exports from the industrial economies were almost four times as large as at the beginning of the 1970s.

There were several reasons for the sharp increase in grain exports from the industrial market economies, but the primary point to be made here is that there was almost no increase in feed use of grains after 1973/1974 in these economies. In fact, the United States has still not regained the level of feed use for 1973/1974, and the increase in feed use in the European Community (EC) and the rest of Western Europe has been very modest. For all industrial market economies the feed use of grain was 280 million metric tons in 1979/1980 or almost identical to the use of 272 million m. tons in 1973/1974 (USDA 1980). As a result, essentially all of the increased grain production in the industrial market economies was available for export or for increasing stocks after 1973/1974. From 1973/1974 until the end of the decade, grain production increased by 66 million m. tons and net exports by 61 million m. tons.

In Eastern Europe and the Soviet Union, growth in the feed use of grain during the 1970s and after 1973/1974 continued. This growth can be attributed to a considerable degree to the food price subsidies that have kept the prices of meat and milk very low, encouraging the consumption of livestock products and the use of grain as feed (Johnson 1982). Had these countries permitted the consumer prices of meat and milk to reflect the prices paid to the farms, instead of subsidizing consumer prices by as much as half, there would have been significantly slower growth in the use of grain as feed. During the 1960s the feed use of grain in the USSR more than doubled (from 43 million m. tons in 1960/1961–1962/1963 to 89 million m. tons in 1969/1970–1971/1972), while in Eastern Europe the increase was slightly more than 60 percent for the same period (from 29 to 46 million metric tons). Grain use continued to increase during the 1970s—by 33 million m. tons or 37 percent in the Soviet Union and by 24 million m. tons or 50 percent in Eastern Europe. The feed use of grain in Eastern Europe and the USSR now exceeds that of the United States by about 60 million m. tons or by 40 percent (USDA 1980).

The feed use of grain is increasing in the developing market economies, though most of the growth is occurring in a small number of middle-income developing economies. These include South Korea, Taiwan,

Brazil, Mexico and the Philippines. But feed use in all of Asia, Africa, and Latin America, excepting only Japan, in 1979/1980 was just 56 million metric tons, or 4 percent of world grain production. Since the late 1960s the annual growth rate in grain used as feed in these economies has been approximately 7 percent. If this growth rate were to continue through the 1980s, feed use of grain at the end of the decade would be about 110 million metric tons or less than 6 percent of the trend level of world grain production for 1989. As rapid as the growth of feed use of grain may be in these economies, continued growth for another decade would not represent a significant threat to the grain availability for people.

Utilization of the World's Potential Cultivatable Land

The view that there remains very little land to bring under cultivation in the world and that this places a significant limitation on the expansion of food production is incorrect. Even if it were true, it would be largely irrelevant. There is a substantial potential for expanding the amount of cultivatable land in South America, Africa, and Southeast Asia (World Food Conference 1974, pp. 82-83). The amount of arable land in the developing market economies had increased during the 1970s, from 565 million hectares in 1970 to 598 million hectares in 1980 or by 5.8 percent (FAO 1980).

But the expansion of the cultivated area is probably not the cheapest means of increasing food and agricultural production. The United States now has the same arable area that it did in 1950, though it is generally agreed that cropland in the United States could be increased by from 10 to 15 percent. However, in increasing crop production at an annual rate of about 2 percent over the past two decades, it has been cheaper for farmers to increase production by improving yields per unit of land than by increasing the area under cultivation.

The developing countries, while continuing to expand the cultivated area, have also been successful in increasing yields. For 1961–1965 to 1978–1980 approximately two-thirds of the 3 percent annual growth in cereal production can be attributed to higher yields (CIMMYT 1981). For that period the growth of grain production in the developing countries was at the same or at a slightly higher rate than in the developed countries.

The emphasis upon the amount of cultivated land ignores the changes that can be made to improve the productive capacity of that land and to significantly increase the stability of production. One of the most important trends in the developing countries has been the increase in the amount of land irrigated. In India the area of irrigated land has increased from 25.5 million hectares in 1961–1965 to 39.1 million hectares in 1979. In China the increase was from 38.5 million to 49.2 million hectares for the same time period. These two countries have a large fraction of the irrigated land in the developing world, where the irrigated area increased from 70 million hectares in 1961–1965 to 100 million hectares in 1979. India alone accounted for almost half of the increase for all developing countries (FAO 1980).

There has also been a significant though smaller increase in irrigated land in the industrial countries—from 27.2 million hectares in 1961–1965 to 31 million hectares in 1979.

Depending upon the particular situations, irrigated land yields two to four times as much as the same land does without irrigation. Thus irrigating one hectare of land is the same as "finding" one to three additional hectares of cropland, even if the irrigated area had been cultivated before.

By the mid-1970s almost all of the land in the United States that had been diverted and idled had returned to cultivation. Not all 24 million hectares diverted under the farm programs during the early 1970s were returned to cultivation, but the acreage of crops harvested increased by approximately 16 million hectares by 1977. Some have argued that returning idle and diverted land to cultivation implies that there is no surplus productive capacity and that the prospects are for slower growth of output in the future than in the past. This conclusion assumes that the acreage diversion programs had a major impact upon agricultural output, yet some work I did several years ago indicated that the effect of the acreage reduction and diversion programs reduced total farm output by 2 percent and crop output by about 4 percent (Johnson 1973). The significant increase in purchased inputs since 1970 has been far more important than the return of land to cultivation in increasing farm production.

Energy Prices

After the sharp increase in petroleum prices in 1973, there was legitimate concern that increased energy prices would have an adverse effect upon the growth of agricultural output in both the industrial and developing economies. Because agriculture in the industrial economies was energy intensive, as were the new yield and output increasing technologies being adopted in the developing economies, it was feared that their benefits would be reduced by higher energy prices. *The Global 2000 Report* gave major emphasis to the effect of sharp increases in energy prices on increases in real food prices.

In particular, it was feared that rapidly increasing fertilizer prices would result in slower yield growth. In the United States, as elsewhere, fertilizer prices increased sharply between 1970 and 1975—by 140 percent

TABLE 3

Farm Input Prices, the United States and the European Community

(1970 = 100)

	1973	1975	1977	1979	1980	1981	Mid-1982
United States							
All Production Items	135	169	185	231	256	274	278
Fertilizer	117	250	208	225	279	300	304
Agricultural Chemicals	108	165	161	154	165	178	195
Fuels and Energy	110	169	192	263	362	410	384
Feed	159	185	185	203	228	248	235
European Community[a]							
Goods and Services Used in Agriculture	130	169	211	234	257	291	--
Fertilizer	120	198	211	243	286	332	--
Energy and Lubricants	123	194	245	306	376	454	--
Feeding Stuffs	134	156	204	215	228	258	--

Sources: United States: U.S. Department of Agriculture, Agricultural Prices, various issues. European Community: Commission of the European Communities, The Agricultural Situation in the Community, annual reports for 1976, 1980 and 1981.

[a]Data are for the nine members except that 1981 price indexes were based on differences between 1981 and 1980 prices for the ten members.

in terms of prices paid by farmers. The nominal price fell by one-sixth over the next two years. Table 3 presents the paid price indexes for U.S. farmers for all production items and for three production inputs with a large energy component—fertilizer, agricultural chemicals, and fuels and energy. Compared to 1970, fertilizer prices at mid-1982 increased by less than 10 percent more than all farm production items. The prices of agricultural chemicals increased by only 70 percent as much as all farm production items. Fuel and energy prices, however, increased by about 40 percent more than all production items.

The U.S. comparisons are affected by the base year chosen. If 1967 were the base year, the mid-1982 index of fertilizer prices would be 12 percent *below* the index of prices for all production items, while fuels and energy prices would have increased by a little less than one-third. It is clear from the table and from the price comparisons with the 1967 base that relative shifts in the prices of production items with a high energy component have not been great. In particular, the increases in fertilizer prices were modest indeed compared to the increase in petroleum prices, from about $3 per barrel in 1970 to more than $35 per barrel in 1981.

The pattern of relative price changes for energy-intensive farm inputs in the European Community during the 1970s was similar to what occurred in the United States. The data are presented in the bottom part of Table 3. Compared to 1970, fertilizer prices in the EC had increased by 14 percent compared to the index of goods and services used in agriculture. The relative index of prices of energy and lubricants increased by 56 percent in the EC and by 50 percent in the United States.

It may be argued that fertilizer prices have not yet felt the full brunt of increased energy costs in the United States because natural gas prices are still under control. Natural gas is the primary energy source for the production of nitrogen fertilizers. When natural gas prices are fully decontrolled, there will be an increase in the cost of producing fertilizer in the United States. But this may result in a significant increase in the price of nitrogen fertilizers in the United States. What is more likely is that production of nitrogen fertilizer will decline in the United States and increase in areas where natural gas continues to be flared or where gas prices are lower than in the United States. Fertilizers are freely traded in the world, and U.S. prices cannot depart greatly from the prices in the international markets.

Prices of energy and energy-intensive products used in agriculture have not increased so dramatically as to impose a major threat to food production. Productivity improvements in fertilizer production have been an important factor in limiting the impact of higher energy prices. In addition, the prices that farmers pay for a particular product reflect many costs other than energy—transportation, marketing services, and credit, for example. Thus it was naive to assume that there would be anything approximating a one-to-one relationship between changes in the price of oil and the products farmers use.

In the developing countries there are numerous governmental interventions that affect the price of fertilizer, and in many countries actual and official prices may differ substantially. However, a review of FAO fertilizer price data for several developing countries gives approximately the same pattern of price changes as occurred in the United States or the European Community, though there is considerable diversity from country to country. In many developing countries the ready and assured availability of fertilizer may be of greater concern than prices.

Prospects for the 1980s

The previous discussion has given important reasons why a sharp increase in the real price of food and other agricultural products did not occur by the late 1970s as had been feared by many. Other factors besides those discussed were also important. In particular, productivity changes continued to be important, providing additional food output while offsetting much if not all of the small increases in the relative prices of energy-intensive farm production inputs. Some feared during the mid-1970s that productivity growth was slowing down in the United States, but they apparently failed to consider the impact of climate upon agricultural output in the United States. Perhaps the most remarkable aspect of agriculture in the United States and Western Europe is that productivity improvements have continued at a rapid rate despite sharp declines in productivity growth in the nonfarm sectors of the economies.

There is, of course, the possibility that the factors that prevented an increase in the real prices of major food products during the 1970s will not operate in the future. The final part of this paper will consider whether there is a reasonable chance that the long-term slow downward trend in the prices of major food products in international markets is likely to be reversed. By reversed I mean a change that lasts for a significant period of time, such as a decade. A short-term reversal is possible and should be expected some time during the next decade. Current governmental policies assure that such a short-term reversal is highly probable. Given those policies, any significant decline in world production that cannot be fully offset by existing stocks will increase grain prices. But such a reversal is likely to have a life similar to the price patterns of the 1970s (Johnson 1975a).

I shall consider briefly four factors that have been put forward as potential sources of real increases in the costs or prices of major food products. These are the use of crops for energy production, the wastage of U.S. land through erosion, water limitations, and increased demand for imports by the centrally planned economies and the middle-income developing economies.

Agricultural Products as an Energy Source

Only two years ago there was an emphasis upon the potential impact of producing large quantities of ethanol from corn and sugar upon the real prices of agricultural products. The United States had announced a program to produce 10 billion gallons (38 billion liters) of ethanol from farm products, primarily corn, by 1990. This program, if carried out, would have required 100 million m. tons of corn or perhaps one-fourth of total grain production in 1990.[2] If such a program had been carried out, there is little doubt that it would have sharply increased the real costs of producing grain and the prices of the major grains. Similarly, if the ambitious Brazilian plans for producing a large share of motor fuel from sugar had been carried out, sugar prices would have moved to a much higher level.

On the basis of economic efficiency, neither the U.S. nor the Brazilian alcohol program made sense. Fortunately, the gasohol program in the United States has been greatly reduced, and the same appears to be true of the Brazilian program. Both programs required large-scale subsidies, and in the case of the U.S. program the net gain in energy that could substitute for fossil fuels was modest indeed. It can now be assumed that for some time in the future our cars and trucks will not be a major threat to the world price of food.

Erosion of Our Soil Resources

In recent years a great deal of attention has been given to the loss of our soil resources. One aspect that was emphasized was the conversion of our agricultural land to nonfarm purposes—for streets, roads, houses, airports, urban commercial development. Such conversion does occur and will continue to occur, though probably at a slower rate in the future than in the recent past due to the decline in population growth and the response to higher energy costs. Although such losses have occurred, their importance was often exaggerated by the repeated claim that 3 million acres (1,215,000 hectares) were being lost each year. For 1967–1975 the annual conversion of cropland to urban expansion and transportation was 245,000 hectares; in addition, water projects covered 30,000 hectares of cropland annually (Crosson 1982). The total of these two losses was 0.17 percent of all cropland; the loss in six years amounted to 1.0 percent of cropland. It will not be too difficult to add new cropland at that rate for some time into the future.

I am in agreement with my colleague T. W. Schultz that the recent statements of alarm about soil erosion are largely misplaced, at least if interpreted to mean that soil erosion has become a more serious threat to the U.S. agricultural productivity than it was several decades ago. In his paper "The Dynamics of Soil Erosion in the United States: A Critical Review" (Schultz 1982), Schultz marshalls direct evidence to support the view that erosion is probably less of a problem today than it was 50 years ago. Even more important are the points made in the following paragraph:

There is an odd ambivalence in assessing the performance of farmers. We proclaim to the world that U.S. farmers are second to none in their agricultural achievements. When it comes to soil erosion, the prevailing implicit assumption is that U.S. farmers have no perception of the value of their soil resources and that they act as if they were indifferent to soil losses. The dynamics and success of U.S. agriculture is ample proof that farmers are competent entrepreneurs. Nor do they shed their entrepreneurial ability when it comes to investments to improve and maintain their soil resources (p. 18).

Soil erosion, whether due to wind or water, can and does have adverse effects upon agricultural output, but it is impossible to eliminate all erosion and uneconomic to eliminate much of the erosion that does occur. As in all things, we could do better. We could, for example, concentrate the efforts of the Soil Conservation Service where erosion is the most serious, rather than spreading such efforts among all or most congressional districts. In our policies and attitudes toward erosion we should start from the position that most of the time farmers know what they are doing and that, at best, governmental programs and policies can have marginal positive consequences.

In a well-documented article, Lester Brown (1981) has assembled a substantial amount of data on the loss of cropland throughout the world. He emphasizes the transfer to nonagricultural uses as well as losses due to erosion and salinity. Although he agrees that total cropland and land devoted to cereals continues to increase, he argues that the new land is less productive than the old. He notes that reasonable projections of the increase in the area devoted to cereals and in population by the year 2000 indicate that the amount of land producing cereals per capita will decline from 0.184 in 1975 to 0.128 in 2000 (p. 85). The expected decline in per capita land base for cereals is not due to a smaller area devoted to cereals but to a projected population increase of 58 percent. It may be noted that the World Bank (1980) projection of world population is for a 52 percent increase from 1975 to 2000, a not insignificant reduction from 58 percent.

The projection of world food scarcity did not arise solely from the decline in cropland per capita in Brown's depiction of the remainder of this century. Two other speculative considerations were involved. The first was the acceptance of projections by FAO and the IFPRI that "world food demand will roughly double between 1975 and 2000" (p. 85). The second was the assertion that the postwar trend of rising yield per hectare "has been arrested or reversed in the United States, France, and China, each the leading cereal producer on its respective continent" (p. 86). The yield statement is false, to put it bluntly. The paper referred to was given in 1980. Yield data for the three countries indicate the following increases for 1977–1980 compared to 1972–1976: France (wheat and coarse grains) 13 percent; United States (wheat and coarse grains)

13 percent; and China (wheat, coarse grains and rice) 15 percent.[3] These are increases that occurred over a period of 4½ years. The data hardly support the view that the rising yield per hectare "has been arrested or reversed." *If* these average yield increases were maintained for 25 years, grain production in these three countries would double, assuming no increase in grain area over the 1972-1976 level. For periods as short as 4 to 5 years weather factors can influence yields; thus I do not believe that the observed yield increases should be used to project long-run yield changes. But I do believe the yield changes are large enough to contradict Brown's statement.

Nor do I believe that one should accept uncritically the FAO and IFPRI projections that the world's demand for food will double in the last quarter of the century. Both population growth and per capita demand may grow at a somewhat lower rate than envisaged in the projections. This seems quite clearly to have been true for both population growth and per capita food demand; the rate of growth of per capita income for the first third of the last 25 years of the century has been lower than generally anticipated. *The Global 2000 Report* projects world food consumption in 2000 at a maximum of 198 percent of 1969-1971 levels (Barney 1982, p. 3). This projection was based on somewhat higher population and income growth rates than now appear probable. Also, the projected increase is for 30 years rather than for 25 years. The difference between consumption doubling in 30 years instead of 25 years is significant, as the longer period would require an annual growth rate of 2.3 percent and the shorter period one of 2.8 percent.

Water Shortages

Many efforts to appraise the future of world food production emphasize the prospect of water shortages or limitations as a significant negative factor. Compared to the gloomy projections in *The Global 2000 Report* with respect to agriculture, the discussion of water availability is almost rosy. It was concluded that "there will apparently be adequate water available on the earth to satisfy aggregate projected withdrawals in the year 2000; the same finding holds for each of the continents. Nevertheless, because of the regional and temporal nature of the water resource, water shortages even before 2000 will probably be more frequent and more severe than those experienced today" (p. 158).

As noted earlier, the developing countries have made substantial investments in the expansion of irrigation during the past two decades, and it is reasonable to assume that this trend will continue for the rest of this century. Most of the factors controlling the distribution and use of water from rivers, lakes, or storage are determined by governments. This means that when water shortages do occur, as was the case in

New York City a few years ago, it is due to a failure on the part of one or more governmental units. Many of the problems currently associated with excess use of irrigation water in the world result from the policies adopted by governments with respect to how such water is allocated and priced. Most irrigation water is underpriced; much irrigation water is free to the user, especially in the centrally planned economies. Under these circumstances there is a great waste of the existing supplies of water, but the potential for saving water is very great if appropriate policies are followed.

Competition for Available Food Supplies

During the 1970s two groups of countries accounted for almost all of the world's increase in grain imports. These were the centrally planned economies and the middle-income market economies. The first group had a population of 1,321 million and the second group 933 million in mid-1978; together they accounted for more than half of the world's population. Between 1969–1971 and 1980, world grain imports increased by 118 million metric tons (see Table 4). Over the same period the net grain imports of the centrally planned economies increased by 56 million m. tons, and the imports of the middle-income developing countries increased by 37 million m. tons. These two groups of countries accounted for 93 million m. tons out of the total increase of 118 million m. tons, or almost four-fifths of the total increase.

Contrary to expectations at the time of the World Food Conference that the demands of the low-income countries, primarily in southern Asia, would result in grain imports of such magnitudes as to put pressure on available export supplies, the low-income countries hardly increased their grain imports at all. Nor did the industrial economies contribute significantly to the import demand for grain. Their gross grain imports increased by 22 million m. tons during the decade, or less than one-fifth of the total increase in grain imports. As noted earlier, the industrial countries increased their gross and net exports by a substantial amount, with the United States alone accounting for almost 70 percent of the total increase in grain exports for the decade.

It is reasonable to expect the middle-income countries to increase their grain imports during the 1980s. However, even if imports were to double, there seems to be little likelihood that this growth would result in an increase in world grain prices.

The centrally planned economies were the major source of growth in world grain trade during the 1970s. Growth has continued into the early 1980s; for 1981/1982 net grain imports had increased to 72 million metric tons, compared to 41 million m. tons for 1977/1979 and 62

TABLE 4.—International Trade in Cereals by Economic Groups, 1960-62, 1969-71 and 1977-79 (million metric tons)

Country Group	1960-62			1969-71			1977-79			1980		
	Export	Import	Net	Export	Import	Net	Export	Import	Net	Export	Import	Net
Industrial Countries	53.8	37.1	16.7	77.6	52.1	25.5	148.4	63.1	85.3	194.3	74.4	119.9
United States	31.4	0.6	30.8	36.3	0.4	35.9	90.9	0.2	90.7	112.9	0.2	112.7
Canada	10.2	0.7	9.5	13.7	0.5	13.2	18.5	0.7	17.8	21.6	1.4	20.2
Australia	5.9	--	5.9	8.8	--	8.8	11.7	--	11.7	19.5	--	19.5
France	3.4	1.0	2.4	11.4	1.0	10.4	14.3	1.9	12.4	19.6	1.6	18.0
Japan	0.1	5.0	-4.9	0.7	14.7	-14.0	0.3	23.3	-23.0	0.8	24.5	-23.7
Centrally Planned	9.8	12.4	-2.6	12.4	17.5	-5.2	9.3	50.0	-40.7	7.7	69.2	-61.5
USSR	7.6	0.6	7.0	8.2	2.7	5.5	3.7	20.7	-17.0	2.3	31.2	-28.9
Eastern Europe	1.3	8.3	-7.0	2.2	9.7	-7.5	4.1	16.6	-12.5	3.6	17.2	-13.6
China	0.9	3.5	-2.6	2.0	5.2	-3.2	1.5	12.7	-11.2	1.4	17.8	-16.4
Low Income Countries	2.4	7.3	-4.9	2.1	10.9	-8.8	2.8	11.7	-8.9	2.8	14.9	-12.1
India	--	4.1	-4.1	--	3.6	-3.6	0.8	0.6	0.2	0.5	0.1	0.4
Indonesia	--	1.2	-1.2	0.2	1.3	-1.1	--	2.8	-2.8	--	3.6	-3.6
Middle Income Countries	9.6	10.9	-1.3	17.6	24.2	-6.6	26.1	49.2	-23.1	17.4	53.1	-35.7
Korea	--	0.5	-0.5	--	2.6	-2.6	--	4.1	-4.1	--	5.1	-5.1
Argentina	5.6	--	5.6	3.5	0.1	3.4	14.6	--	14.6	10.0	--	10.0
Brazil	0.1	2.1	-2.0	1.2	2.1	0.9	0.7	4.8	-4.1	--	6.7	-6.7
Mexico	0.2	0.1	+0.1	0.5	0.4	0.1	0.1	4.0	-3.9	--	7.1	-7.1
South Africa	1.3	0.2	1.1	0.3	1.2	0.9	2.6	0.2	2.4	3.8	0.2	3.6
Thailand	1.9	--	1.9	2.9	0.1	2.8	4.4	0.1	4.3	5.1	0.2	4.9
Capital Surplus Oil Exporters	--	0.7	-0.7	0.1	2.0	-1.9	0.1	6.8	-6.7	0.3	10.3	-10.0
Total	75.6	68.4	--	109.8	103.9	--	186.9	183.3	--	222.5	221.9	--

Source: FAO, *FAO Trade Yearbook*, various issues.

million m. tons in 1980. All of the recent growth has been due to Soviet imports.

The high level of Soviet grain imports reflects the effects of recent unfavorable weather, generally slow output growth prior to the recent run of poor grain crops, and a policy of holding constant the nominal prices for livestock products for the past two decades. Even with a return to long-run average climatic conditions, the Soviet Union is likely to continue to be a major grain importer, perhaps importing 30 to 40 million m. tons annually. Such large imports will be required if there is to be any significant increase in Soviet per capita meat consumption, which has not increased since 1975 and remains near the lowest level in Eastern Europe. Pressure to increase meat production exists, in part, because of the low subsidized prices of meat, which cause the gap between demand and supply to grow year by year.

The Eastern European countries increased their annual grain imports from a little less than 10 million metric tons in 1969–1971 to 17 million m. tons in 1980. Grain imports by these countries are now significantly below the 1979/1980 level of 17.5 million m. tons; 1981/1982 grain imports were less than 14 million m. tons, and a further reduction in imports is projected for 1982/1983. The primary reason for the recent reductions in grain imports is the lack of foreign exchange required to pay for the imports. The recent events in Poland are responsible for a significant part of the decline, but other economies have also had to restrict grain imports.

There is uncertainty concerning Chinese intentions. Recent levels of grain imports—about 15 million metric tons—could be maintained for the future or there could be a decline or increase. A reasonable case can be made that grain imports will not decline. In the early 1980s 40 percent of the grain consumed in cities was imported. To cut back significantly on this import dependence would be difficult, at least for the next few years. However, I do not see any reasonable basis for projecting a major increase in Chinese grain imports. The present level of imports already places a significant burden upon the available foreign exchange. Barring a quite unsatisfactory performance of Chinese agriculture, an increase in grain imports to more than 20 million m. tons during the decade would come as a pleasant surprise for North American grain growers.

There is no reasonable expectation that the grain imports of the centrally planned economies can increase during the 1980s at the same absolute rate as during the 1970s. During 1981/1982 these economies had net grain imports of 72 million metric tons, or about one-third of world grain imports. Grain imports in excess of 80 million m. tons annually seem quite unlikely; a decline to 60 million m. tons would not come as too great a surprise.

I consider the 1980s in a pessimistic manner if they are viewed in the context of the demand for U.S. agricultural products. The demand for U.S. agricultural exports is likely to grow much more slowly during the 1980s than during the 1970s. O'Brien (1981) has projected export growth for the first half of the 1980s at less than one-third of the actual growth for the 1970s. From 1972–1980 the annual rates of growth in the volume of major farm exports were 12 percent for grains, 10 percent for oilseeds and products, and 16 percent for cotton. For 1979/1980 to 1985/1986, annual growth rates were projected at 3 percent for feed grains, 2 percent for wheat, and 1.5 percent for soybeans. Cotton exports were projected to stay at the same level in 1985/1986 as the average of 1977/1978–1979/1980 (pp. 14, 13). I see no basis for projecting more rapid growth rates than these for the entire 1980s. The projected slow growth of U.S. agricultural exports, with similar slow growth rates for the other temperate-zone agricultural exporters, means that there will not be real price increases for the major food products. Instead, the probable outcome will be downward pressure on prices throughout most of the 1980s.

Conclusions

Although I believe that there has been a modest improvement in food and nutrition in the poor countries of the world (with the possible exception of a number of African countries), I am confident that the rate of improvement can be increased if further efforts are made. The recent rate of improvement has resulted from a number of conscious efforts on the parts of governments and from the application of the enormous talents and initiative of hundreds of millions of farm families. The output growth has resulted from agricultural research, investments in irrigation, development of roads, the expansion of markets, increases in education and literacy of farm people, the availability of extension services, and increased ease of communication. Further expenditures and investments in these areas will yield additional returns.

Much can be done to improve the economic setting within which farmers function. Policies that assist farmers must replace policies that put barriers in the way of farmers and other poor people in their efforts to improve their incomes and the satisfactions that they derive from life. Although there appears to have been some improvement during the past decade, all too many countries still exploit farm people through export taxes and other methods of keeping domestic prices below world market levels.

Notes

1. The two scenarios, labeled A and B, are not forecasts: "They are quantified explorations of feasible future developments." Scenario A assumes relatively rapid growth of gross domestic product for 90 developing countries of 7.0 percent, while Scenario B assumes a 5.7 percent annual growth rate in GDP. Each scenario develops growth patterns for food and agricultural production and changes in nutritional status. The factors that might lead to the differences in the rate of growth of agricultural production are outlined in Chapter 4 of FAO, *Agriculture: Toward 2000* (Rome: FAO, 1981).

2. If 100 million metric tons of corn were used to produce ethanol, approximately 30 million m. tons of a medium-level protein feed would be available. This 30 million m. tons of feed would replace about 18 million m. tons of soybean meal.

3. Two points may be made concerning yield changes. First, my comparison of 1977–1980 yields with those of 1972–1976 was a matter of convenience; in the source used the average yield for 1972–1976 was the earliest period given. The average U.S. yield of wheat and coarse grains together for 1972–1976 at 3.5 m. tons per hectare was slightly below the 1969–1973 average of 3.56. It was perhaps this small dip in yields that convinced Brown that grain yields had peaked. Second, national grain production is not determined solely by the yields of individual crops but is influenced by the choice of crops. In the United States there has been a shift among grain crops toward the higher-yielding crops, namely corn and wheat, and away from barley and oats over the past two decades. Data from Foreign Agricultural Service, U.S. Department of Agriculture, foreign agriculture circular, FG-2-79, Jan. 2, 1979, pp. 3–4, and FG-36-80, Dec. 2, 1980, pp. 2–3.

References

Barney, Gerald O., study director. *The Global 2000 Report to the President.* Prepared by the Council on Environmental Policy and the U.S. Department of State. New York: Penguin Books, 1982.

Brown, Lester R. "The Worldwide Loss of Cropland." In *Future Dimensions of World Food and Population,* ed. Richard G. Woods. Boulder, Colo.: Westview Press, 1981, pp. 57–96.

Crosson, Pierre R., ed. *The Cropland Crisis: Myth or Reality?* Washington, D.C.: Resources for the Future, 1982.

Food and Agriculture Organization (FAO). *FAO Production Yearbook.* Rome: FAO, annual issues.

————. *FAO Trade Yearbook.* Rome: FAO, annual issues.

————. *Agriculture: Toward 2000.* Rome: FAO, 1981.

International Food Policy Research Institute (IFPRI). *Report 1981.* Washington, D.C.: IFPRI, 1982.

International Maize and Wheat Improvement Center (CIMMYT). *World Wheat Facts and Trends.* Report One. Mexico: CIMMYT, August 1981.

Johnson, D. Gale. *Farm Commodity Programs: An Opportunity for Change.* Washington, D.C.: American Enterprise Institute for Public Policy Research, 1973.

————. "World Agriculture, Commodity Policy, and Price Variability." *Am. J. Agri. Econ.* 57 (1975a): 823–828.

————. *World Food Problems and Prospects.* Washington, D.C.: American Enterprise Institute for Public Policy Research, 1975b.

————. "Inflation, Agricultural Output, and Productivity." *Am. J. Agr. Econ.* 62 (1980): 917–923.

————. "Agriculture in the Centrally Planned Economies." Chicago: Office of Agricultural Economics Research, University of Chicago, Paper No. 82-15, 21 July 1982.

O'Brien, Patrick M. "Global Prospects for Agriculture." In *Agricultural-Food Policy Review: Perspectives for the 1980s* AFPR-4, April 1981, pp. 2–26.

Schultz, T. W. "The Dynamics of Soil Erosion in the United States: A Critical Review." Chicago: Office of Agricultural Economics Research, University of Chicago, Paper No. 82-9, 12 March 1982; revised 22 March 1982.

U.S. Department of Agriculture. Foreign Agricultural Service. *Utilization of Grain for Livestock Feed.* Foreign agricultural circular, grains, FG-14-80, 1 May 1980. Washington, D.C.: USDA.

————. Economic Research Service. *World Indices of Agricultural and Food Production.* Stat. Bul. No. 669, July 1981. Washington, D.C.: USDA.

Valdés, Alberto, and Joachim Zietz. *Agricultural Protection in OECD Countries: Its Cost to Less-Developed Countries.* Research Report 21. Washington, D.C.: IFPRI, 1980.

World Bank. *World Development Report.* Washington, D.C.: World Bank, annual issues.

World Food Conference. *Preliminary Assessment of the World Food Situation: Present and Future.* E/Conf. 65/3. Rome: FAO, 1974, Chapter 5, paragraph 281-284.

Discussion _____

Raymond F. Hopkins
Department of Political Science
Swarthmore College

D. Gale Johnson has prepared a lucid and persuasive assessment of the current world food situation. I am flattered that he invited me, a political scientist, to comment on his paper. I presume from this that he wanted some controversy; I shall try not to be disappointing, following the basic premise that more is gained from exploring disagreements than points of agreement.

Although most of my comments differ from Johnson's interpretations, I find myself more in agreement than disagreement with his paper. In particular, I agree that prospects for food shortages and high prices *were* exaggerated during the panic market period of 1973–1975, and they continue to be exaggerated by some scholars. Furthermore, trends have been favorable since the 1950s in per capita food production and grain trade. These facts are unassailable, as is Professor Johnson's conclusion that the growth in grain imports during the 1970s is principally attributable to socialist and newly industrializing countries (NICs), not the poorest developing countries. Finally, I admire the way the paper takes to task some of the forecasts of tight food supplies by the year 2000. In addition to the specific points Professor Johnson makes, there is good reason to be extremely skeptical of such long-range forecasts in general. More often than not such forecasts prove to be grossly in error, as studies of past forecasting have shown; policy-makers are well advised to treat such long-range forecasts with extreme skepticism.[1] The shift in focus in the paper to the decade of the 1980s is, I believe, about right for informed judgment. Regrettably, few policy decisions rest upon considerations of even this more reasonable future concern; rather, most policies are driven by something between last year's conditions and an eighteen-month forward expectation.

Let me turn now to some criticisms. There is quite a dark side to Johnson's characterization of the last few decades as "things getting better." I also disagree with his assessment that demand growth will be very slow and prices low for the 1980s. I think this forecast will prove accurate only if the current worldwide recession continues. Finally, and not surprisingly, I see the role of government intervention into markets somewhat differently. For me the question is not whether government intervention is good or bad, but rather how it can be improved in politically realistic ways.

The Dark Side of the "Things Are Improving Slowly" Scenario

Professor Johnson argues that in spite of his cautiously optimistic assessment of factors affecting hunger and undernutrition, "there is no room for complacency." This is certainly the case.[2] We must be especially concerned because two trends other than per capita production are relevant. First, most people compare themselves with one another and compare their situation to their immediate past rather than with that of the previous generation. Thus for the 20 to 40 percent of the world's populace that are least well fed, a short-term deterioration in their condition or an awareness of growing disparity between themselves and better-fed populations might trigger serious economic and political disturbances. At a minimum, short-term threats to those lacking personal sources of reserves and resentment over inequality will increase pressure for governments to intervene in markets through subsidy programs. Second, many people compare existing food production and distribution activities with known potential. Professor Johnson himself goes to some length to assure us that the land, water, transportation, improved education, and other resources for greater production exist. Is it possible that we are falling behind in realizing these possibilities? If the ratio of actual to potential world production and nutrition declines—which could occur if barriers to innovation and distribution multiply—then the apparent modest gains must be seen as especially inadequate.

The forecasts seen as cautiously optimistic by Johnson look dangerous to me. The FAO trend forecast projecting a decrease in undernutrition from 23 to 17 percent between 1975 and 2000 actually implies an increase of 100 million additional seriously underfed people.[3] The FAO trends, compared to those of IFPRI or Lester Brown, are the most promising ones cited. Everyone agrees that much more can and should be done to improve trend forecasts. The issue is: How bad would things be if trends persisted because either improvements are not undertaken or other countervailing forces arise? The answer is quite bad, especially in sub-Saharan Africa with its extremely high birth rates.

Africa is the darkest cloud on the food horizon. Professor Johnson indicated that "Africa had a constant average level of per capita food production in the 1950s and 1960s and a shocking decline during the 1970s" (p. 8). In his estimate, 1980 per capita production was 15 percent below 1969–1971 levels, though he doubts this figure based on gains in life expectancy. In any event, compared to Asia and the rest of the world Africa is the most acute if not the largest problem for the current world food system.

Others find the African case, now and for the near future, quite grim. In his ministerial review of food in March 1982, Maurice Williams of the World Food Council claimed: "Analysis of the African food problem reveals a steady deterioration in African food production, far outstripped by population growth. The decline in food production per person was 7 percent in the 1960s accelerating downward with a decline of 15 percent in the 1970s. The outlook for the 1980s is grim, and hunger and malnutrition during the 1980s can be expected to become far more widespread."[4]

Our focus is further sharpened by examining wheat trade rather than grain trade as Professor Johnson does. Most of the wheat traded goes directly to human consumption, while most coarse grain is used for feed. The growth in imports for many less-developed countries (LDCs) was dramatic in wheat; the impact on trade and potential food dependency is probably greater than the grain figures shown in Johnson's Table 4 (p. 29) suggest. Sub-Saharan African countries doubled their wheat imports between 1974/1975 and 1981/1982; their growth was about the same as the Soviet Union's and China's and larger than any other region's. Africa, including Egypt, accounted for over 16 percent of wheat imports in 1980/1981, compared to 4 percent for Eastern Europe, 13 percent for China, and 19 percent for the USSR.[5] I believe LDCs will be especially important sources for growth in wheat trade, as well as grains. Regardless of how Johnson or others divide the LDCs into various subgroups, such as NICs, oil-exporters, etc. (and I think China should be included as an LDC), these are the areas of most rapid urbanization and highest income growth, and hence the major source for expanded grain trade in the 1980s. I expect Professor Johnson agrees. Our major differences would be over (1) how tight world markets may become given no expected growth in Soviet imports, and (2) whether growth in LDC imports can continue in the 1980s as it did in the 1970s. The answer to both questions, I believe, lies outside the grain markets. Simply extrapolating population growth, shifts in tastes due to affluence and urbanization, and various production constraints in

LDCs suggests strong latent demand. If the current world recession ends soon—a recession that already may be blamed for the first decline in U.S. agricultural exports in a decade—I expect LDC demand growth to surpass that of other importers. Pressure for subsidies should increase, so that a stable portion of these LDC imports will be subsidized, especially those for bankrupt countries in Africa and the Middle East.

The Depressed Markets Scenario

This last point leads me to another area of disagreement. Professor Johnson sees "downward pressure on prices [in world grain markets] throughout most of the 1980s" (see p. 31). The world food market is closely related to the world economic situation. Grain prices were depressed in the 1930s not because of high yields, but thanks primarily to the world depression; whether trade barriers were the fundamental or more coincidental cause of this I shall not say. I am confident, however, that with prices as low as they are today, only great gains in productive efficiency are saving from bankruptcy those farmers in the United States and elsewhere whose incomes are tied to world prices. Neither soil erosion nor water shortages should threaten such gains in the coming decade. In this longer term with a general recovery in the international economy, demand will reassert itself. The resources that can allow production to grow will not be free, however. Longer-term higher prices for food and water go together in the improved water use future that Johnson expects, for example. In the near term, government measures such as setting acreage aside are already combining to lower production. Hence, I foresee a firming of world prices over the next year or two and upward pressure in real terms for the last half of the decade, provided there is a general economic recovery in the Western world.

Markets and Governments

Weak governments are bad. Most people want strong governments and want them to intervene in markets, at least in research and innovation markets or the world nuclear weapons market. Food markets are no different. People want their government to guard against "excessive" risk and regulate free riders from capturing "unfair" benefits. Government intervention into food markets is extensive, lengthy, and not without bad effects.[6] Yet I think in his paper Professor Johnson lays too diffuse a blame on governments for price instability and barriers to production. Of course, he does not criticize governments generally. He finds that some public goods provided by governments, such as education, information, roads, and extension services, are desirable. From his writings I know he has advocated that governments of wealthy countries should play an insurance role to help stabilize poor countries' food supplies. I do not understand, therefore, why the paper lays so much of the blame for erratic food prices and barriers to production on governments, particularly government "intervention." Less government intervention is not a panacea.

I am not arguing that governments do the right thing even most of the time. In Africa, bad government policies are certainly to blame for a great deal of the continent's food problems.[7] The task is to work to improve the role of government in structuring optimal conditions for markets to operate in. Probably all observers agree that governments help food markets by providing information on crop conditions, by supporting research, by enforcing contracts, including futures contracts, and so on. Disagreement occurs over how much risk governments should cover and what mechanisms they should use. It is not clear that the correlation Professor Johnson finds between greater government regulation and price instability for five major traded commodities indicates a causal relationship. Surely if we found that governments sent more fire engines to fires that did the greatest damage, we would not blame the arrival of the fire engines for the losses, nor prescribe their elimination. There is, in short, usually a good reason for government intervention into social life, and it usually is motivated by a desire to improve the life circumstances of people by reducing or sharing risks.

To be fair, I do agree with Professor Johnson that governments can distort markets. Europe with its common agricultural policy (CAP) and the Soviet Union with its trading and price policies have extended some food security and price stability to their populations at a cost of greater instability in world markets, just as he has argued. It is correct to assign blame to such government policies that thin the world market in times of stress as in 1973–1974. These government policies contributed to international price instability. The answer is to adjust the CAP, however; to advocate abolishing it is simply not practical. There are other government intervention policies, national and international, that I believe make a positive contribution to the performance of world food markets, especially given the other policies that exist. For example, the food security for LDCs and the stability of their export demand are helped by the newly created International Monetary Fund (IMF) loan facility for food imports, by some portion of current food aid programs, and by commercial loans under banking conditions facilitated by governments. During the current period of low prices, LDCs might even rationally use U.S. government–secured futures markets to hedge against future needs and price increases. It would be cheaper than storage in an LDC reserve system.[8]

In conclusion, the issue for the 1980s is not whether markets would better serve the world food economy with less or more government intervention. Intervention is here to stay. The role of government must include providing the confidence and rules to create markets and help them function well in adjusting supply and demand conditions.[9] Intervention will also include practices to protect farmers from a prisoner dilemma type of ruin. When such policies are shaped, however, the problem becomes one of preventing excessive regulation, distortion, black markets and so forth on the one hand, and on the other hand of avoiding an inadequate government role. Domestic protection produces external effects at the international level, so that weak countries and producers are faced with either costly policies of self-reliance or excessive risk and adjustment costs from their exposure to excessively unstable markets. Politics dictates that before domestic stability or income policies are altered, however, incentives and assurances will be needed to induce the alterations. Whether for farmers in Africa or consumers in the United States, the issue is what kinds of government intervention would better serve their goals through some balance of efficient production, equitable distribution and security of supply.

Notes

1. See William Ascher, *Forecasting: An Appraisal for Policy Makers and Planners* (Baltimore: Johns Hopkins University Press, 1978).

2. Alan Berg's book, *The Nutrition Factor* (Washington: Brookings Institution, 1973), still ranks as one of the best accounts of the social, moral, and economic debilitations that accompany malnutrition.

3. I derived this estimate using the conservative population estimate of 4 billion in 1975 and 6 billion in 2000 and the FAO figures cited by Johnson (see p. 7 this book).

4. "Food Policy Issues for the Eighth Session," Report by the Executive Director (Rome: World Food Council, 12 March 1982), p. 11. I do not know why Johnson's figure for the 1960s in Africa is a modest gain, while Williams' report indicates there was a 7 percent per capita decline.

5. Food and Agriculture Organization, *Food Outlook* (Rome: FAO, June 1976 and May 1982).

6. A good critique of government intervention is found in Bruce Gardner, *The Governing of Agriculture* (Lawrence: Regents Press of Kansas, 1981).

7. See, for example, the paper by Robert Bates, "Governments and Agricultural Markets in Africa," p. 153 this book.

8. I am indebted to Anne Peck for convincing me of the soundness of this use of markets. See Peck, "Futures Markets, Food Imports and Food Security" (Washington: World Bank Economics and Policy Division, September 1982).

9. See Charles E. Lindblum, *Politics and Markets* (New York: Basic Books, 1977).

Part 2 _____

2
World Hunger:
Extent, Causes, and Cures

Thomas T. Poleman
Department of Agricultural Economics
Cornell University

Let me make clear at the outset that there is no way to specify with certainty the extent of world hunger. To do so would require much more information than is presently at hand about the actual availability of food, the exact amount of food people need for proper nourishment, and how access to food varies among different income groups within a country. These limitations notwithstanding, I would hazard the guess that at no time in history has the world been as well fed as it is today. To argue otherwise would be to deny a basis for the increase in life expectancy—from about 40 to 55 years—that has occurred during the past 30 years in the poorer countries. I would also guess that the principal nutrition problem in the world today is not hunger among the poor, but the result of overeating and underexercising by the affluent in the industrialized countries.

By so saying I do not want to suggest that hunger and its consequences have all but disappeared. Rather it is to suggest that the problem is a manageable one and given proper governmental will and direction could be resolved in almost all parts of the world during the remaining years of the century.

Why then does the conventional wisdom come down so heavily on the other side? Ask the man in the street for his impression of what life is like in the developing countries and he will invariably reply "hungry." This perception finds extensive support in the literature. We were, after all, told 30 years ago by Lord Boyd-Orr (1950, p. 11) the first director general of the United Nations Food and Agriculture Organization (FAO), that "a lifetime of malnutrition and actual hunger is the lot of at least two-thirds of mankind." We were told by one of the many alarmist books that captured the public's attention in the mid-1960s that widespread famine would overtake us by 1975, and just two years ago the Carter administration's Presidential Commission on World Hunger (1980, p. 182) informed us that the "world hunger problem is getting worse rather than better. There are more hungry people than ever before."

Measuring Hunger: The Early Assessments

The notion that the world is no longer able to feed itself and teeters uncontrollably on the brink of mass starvation may be traced to the formative years of the FAO and its early attempts to assess the extent of world hunger. These studies, and their methodological underpinnings, are worth reviewing because, although they are now discredited, their message has remained remarkably durable. This is because few who have seen fit to make pronouncements on world hunger—and this includes the recent presidential commission—have troubled to undertake original research of their own.

The findings of the early FAO studies and those of more recent work by FAO, the U.S. Department of Agriculture (USDA), and the World Bank are shown in Figure 1. The extent of disagreement among them is extraordinary. Boyd-Orr's conclusion that two-thirds of mankind were hungry came from the FAO's *Second World Food Survey* (1952). An earlier survey (1946) suggested a lower figure. The USDA's two *World Food Budgets* (1961, 1964) concluded that almost the entire population of the developing world lived in "diet deficit" countries. FAO's *Third World Food Survey* (1963) put the afflicted in such countries at about 60 percent and identified a shortage of protein as the principal problem. The World Bank (Reutlinger and Selowsky 1976) saw a problem of roughly the same magnitude—involving about 1.2 billion people—but suggested the prime cause was shortfall of calories. In its two most recent studies (1974, 1977) the FAO also identified calories as the culprit, but put the total afflicted at something over 400 million.

The analytical approach followed in FAO's first two *World Food Surveys* and the two *World Food Budgets* prepared by USDA was extremely simple and may be summarized by the following equation:

FIGURE 1. PERSONS IDENTIFIED AS NUTRITIONALLY DEFICIENT IN
IN MAJOR WORLD FOOD ASSESSMENTS

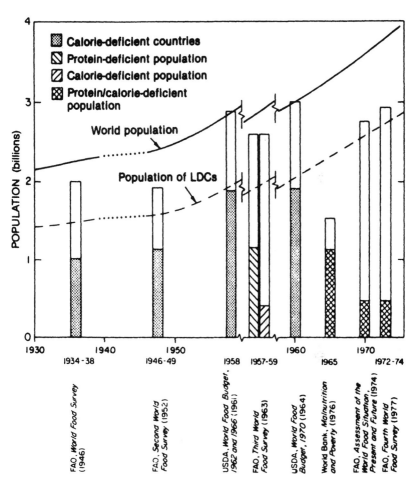

The size of the populations to which the assessments apply is
indicated by the total height of bars.

$$\frac{\text{Food available for human consumption}}{365 \times \text{population}} - \frac{\text{Loss}}{\text{allowance}} \leq \frac{\text{Average daily recommended}}{\text{nutrient allowances}}$$

To determine whether or not a country was experiencing a food problem,
apparent per capita food availabilities, minus an allowance for wastage

between the so-called retail level and actual ingestion, were set against estimates of per capita nutrient needs. Where and when availabilities exceeded requirements, all was presumed well; where they did not, the country or region's entire population was considered to be inadequately nourished.

The failings of this approach are several and, when probed, obvious. First, it assumes that societies are sufficiently homogeneous in their food habits for average data to have meaning. This is certainly not the case in developed economies, where differences in income, locality, ethnic background, and place within the household have long been known to have marked effects on food behavior; this is now recognized to be equally valid for the developing world.

A further drawback of the approach is that it presupposed an ability to specify average food availabilities and needs with a fair degree of precision.

Problems of Estimating Food Availabilities

To estimate food availabilities one must construct a balance sheet, incorporating on the supply side measurements of production, trade, and stocks changes, and on the utilization side such items as seed and feed use and losses in storage. Availabilities for human consumption are derived as a residual and thus reflect the totality of error.

The error so introduced will almost invariably be in the direction of understatement. Understatement of production is a characteristic of most newly developed agricultural reporting systems. Wheat production in the United States is now recognized to have been 30 to 40 percent above that officially reported during the first decade (1866–1875) of USDA's statistical efforts. In Mexico, the extent of understatement for maize during 1925–1934, the *Dirección General de Economía Rural*'s first decade, was over 50 percent.

To this understandable tendency further complications can be added:

- The statistical officer in developing countries is frequently (and not irrationally) equated with the tax collector by the farmer, whose response will be to minimize production.
- Output that is not seen is not counted, and if communications are poor a great deal is not seen.
- Much food production is for on-farm consumption and does not pass through commercial channels where it might be monitored.
- In tropical areas especially, many food crops are not grown in pure stands but mix-planted in fields of bewildering complexity.

To generalize about the extent to which food availabilities in LDCs have been and are now understated is not easy. A reasonable assumption is that the accuracy of production estimates has improved with time and that the extent of understatement is now less than it was when FAO published its first *World Food Survey*. An exception may be sub-Saharan Africa, where independence has frequently been accompanied by a deterioration in the reporting systems established by colonial administrators. When perfection may be anticipated is anybody's guess. It was not until 1902, 36 years after the effort began, that USDA began reporting wheat output with an acceptable margin of error; and not until the mid-1950s, with 30 years of experience in hand, was Mexico able to confidently measure its maize harvest.

Detailed studies of the food economies of Malaysia and Ceylon (Sri Lanka) carried out by my students in the 1960s suggested that caloric availabilities in both were officially understated by between 10 and 15 percent. As the staple in both countries is rice grown under irrigated conditions, which is relatively simple to quantify, and as both countries have by the standards of the developing world an admirable statistical tradition, this 10–15 percent is probably something of a minimum understatement. Elsewhere the amount of food actually available may be undercounted by rather more.

Problems of Estimating Food Requirements

Compounding this tendency to undercount food availabilities have been the difficulties associated with estimating food needs. Until recently these have been overstated. Nutrition is still a young science and our ability to establish minimal or desirable levels of intake is not nearly as precise as we would like it to be.

A person's nutritional needs are a function of many things: age, sex, body size, activity patterns, health status, and individual makeup are the more important. Conceptually, knowing these variables, it should be possible to set minimum levels of intake for protein, energy, vitamins, and other nutrients sufficient to preclude overt deficiency disease in most of a population. Practically it is not possible, and what were used as surrogates for such minimal criteria in the early food evaluations were the recommended allowances prepared as guidelines for dietitians and other nutritional workers. These allowances consciously err on the side of caution, both to incorporate a comfortable safety margin and to ensure that the substantial variations in food needs among individuals will be covered.

The recommended allowances are periodically modified and from the direction and magnitude of change it is possible to infer something of the probable extent by which minimum needs were overstated in the past. With respect to energy allowances, the history of the FAO, the United States Food and Nutrition Board, and other responsible organizations has been one of continued downward modification. The energy allowance for the U.S. "reference man"—in his twenties, weighing 70 kilograms, and not very active—now stands at 2,700 calories, 500 calories less than the 1953 recommendation.

Apart from undue initial conservatism, the principal cause of this reduction is the increasingly inactive character of life in industrial societies. Physical effort is less and less demanded on the job and the body moves from place to place less on its feet than on its seat. It is not unlikely that the energy allowances suggested for the developed countries are now quite reasonable. Little remains to be understood of how urban man divides his day—it has become after all depressingly routinized—and, thanks to studies carried out in association with wartime rationing programs in the United Kingdom and Germany, the energy costs of most activities are well known.

The same is not true for developing countries. Very few energy expenditure/activity studies have been conducted among rural or urban people in these regions and useful common denominators continue to be wanting. A key reason for this shortcoming is the difficulty of obtaining reliable information on energy expenditure. The traditional method for doing this is to record the energy costs of specific tasks with a respirometer and then multiply the resulting factors by appropriate time spans. The problems are many. The respirometer is a clumsy instrument; it can be kept on a subject for only a few minutes and its presence is hardly conducive to normal behavior. Moreover, time span recording must be meticulously accurate in order to be useful. To obtain such information under primitive conditions without an impetus similar to wartime rationing is probably asking too much of the research priorities of most less-developed countries (LDCs).

In response to criticism that the energy allowances used in its early world food surveys were unrealistic surrogates for minimum needs, the FAO, lacking actual evidence from the LDCs, employed a different approach to establishing floor criteria, beginning with its report to the 1974 World Food Conference. This was to estimate the minimal energy requirements of the average nonfasting male at 1.5 times his basal metabolic rate (BMR) and to assume that some individuals might have a BMR as much as 20 percent below the norm.

It is difficult to fault this modification. Certainly the 1.3 BMR factor which results—0.8 × 1.5 BMR—yields values that bear a clearer hallmark of reality. If anything, it errs on the side of being too low. Applied to Asia it suggests minimum per capita requirements of the order of 1,500 calories, as opposed to the criteria of 2,600 and 2,230–2,300 calories, respectively, used in the first two world food surveys.

In the early FAO and USDA studies, the terms "undernourishment" ("undernutrition") and "malnourishment" ("malnutrition") were widely used. Undernourishment is generally taken to mean a shortfall in caloric intake such that a person cannot maintain normal bodily activity without losing weight. Malnourishment, on the other hand, describes the lack or deficiency of one or more of the "protective" nutrients—protein, vitamins, and minerals.

The first two world food surveys defined the nutritional problems of the LDCs largely in terms of energy shortfalls and undernourishment. In the *Third World Food Survey* insufficient protein availabilities and malnourishment were highlighted. Today most nutritionists concerned with the developing countries speak of protein-calorie (or protein-energy) malnutrition. This again identifies a shortage of energy as the prime problem, taking into account that an apparent adequacy of protein can be converted into a deficit should a portion of it be metabolized to compensate for insufficient energy intake. This major change in problem perception and terminology coincided with a drastic reduction in the recommended minimum allowances for protein.

As with energy allowances, the early FAO protein recommendation contained a comfortable safety factor as well as an allowance to take individual variation into account. In 1971 an expert panel concluded that these had been excessive and reduced the daily per capita recommendation for adults by one-third: from 61 grams of reference protein to 40.

The effect of this change was dramatic. Prior to the revision, simple comparisons of average availabilities and needs suggested that almost all the world's developing countries were deficient in protein; after it, hardly any of them. If the "protein gap" did not disappear overnight, its statistical underpinnings seemed to.

Misleading Conclusions

Since they used food availability estimates that *understated* to compare against food requirement figures that *overstated*, it is not surprising that the early FAO and USDA global food assessments painted a gloomy picture of world hunger. Though the numbers varied, the picture was one of hungry countries and of a world unable to feed its rapidly growing population. Insufficient production was seen as *the* problem. As the second of the USDA's *World Food Budgets* (1964, pp. iii-iv) put it:

> Two-thirds of the world's people live in countries with nutritionally inadequate national average diets . . . The diet-deficit countries are poor

and food deficiencies merely reflect the low level of income in general. ... The basic problem of the diet-deficit countries is one of productivity. The people cannot produce enough food to feed themselves or produce enough other products to buy the food they require. Food production has barely been able to keep ahead of population growth, much less provide for the expanded demand resulting from some improvement in per capita income, most of which goes for food.

We now know that such conclusions seriously distort reality. The record of agricultural productivity in the LDCs has not been all that bad. According to such generally used series of "world" food output as that currently issued by the USDA (Figure 2), the LDCs over the past 30 years have expanded production rather more rapidly than the developed countries, a remarkable achievement in view of the minimum priority given to agriculture in their development programs. Population growth, to be sure, absorbed most of the gains, but with the exception of sub-Saharan Africa modest per capita improvement occurred.

There have, of course, been year-to-year fluctuations in this trend—fluctuations whose import has tended to be magnified by those who would influence public opinion. The first apparent faltering came in the mid-1960s and resulted almost exclusively from two successive droughts in India. Indian production bulks so large in the LDC aggregate that major fluctuations in her harvest influence visibly the index for all developing countries. This fact, however, was lost on many commentators. Conditioned by the early FAO and USDA findings to think of all LDCs as "hungry" and hearing of massive food aid shipments (of the 30 million metric tons of grain shipped by the United States under Public Law 480 during the two years ending in June 1967 half went to India) not a few were inclined to predict imminent global starvation. The Paddock brothers (1967) went so far as to specify 1975 as the year in which this would take place.

A reaction set in almost immediately and again closely mirrored the Indian situation. A sequence of favorable years in terms of weather was accompanied by introduction into the Punjab of high-yielding varieties of Mexican wheat. As a result the index for all low-income countries rose steeply, as did per capita availabilities. The assessment was as extreme in the opposite direction as it has been in 1965 and 1966. These were the years when people talked of the Green Revolution. The situation in northwestern India, together with the introduction of high-yielding, fertilizer-responsive rice in the wetter portions of Asia, led many to believe that a fundamental change had taken place and that feeding the world's rapidly increasing population no longer posed problems. So pervasive was this optimism that the FAO went so far as to

FIGURE 2. INDICES OF TOTAL AND PER CAPITA
FOOD PRODUCTION, 1951-1980

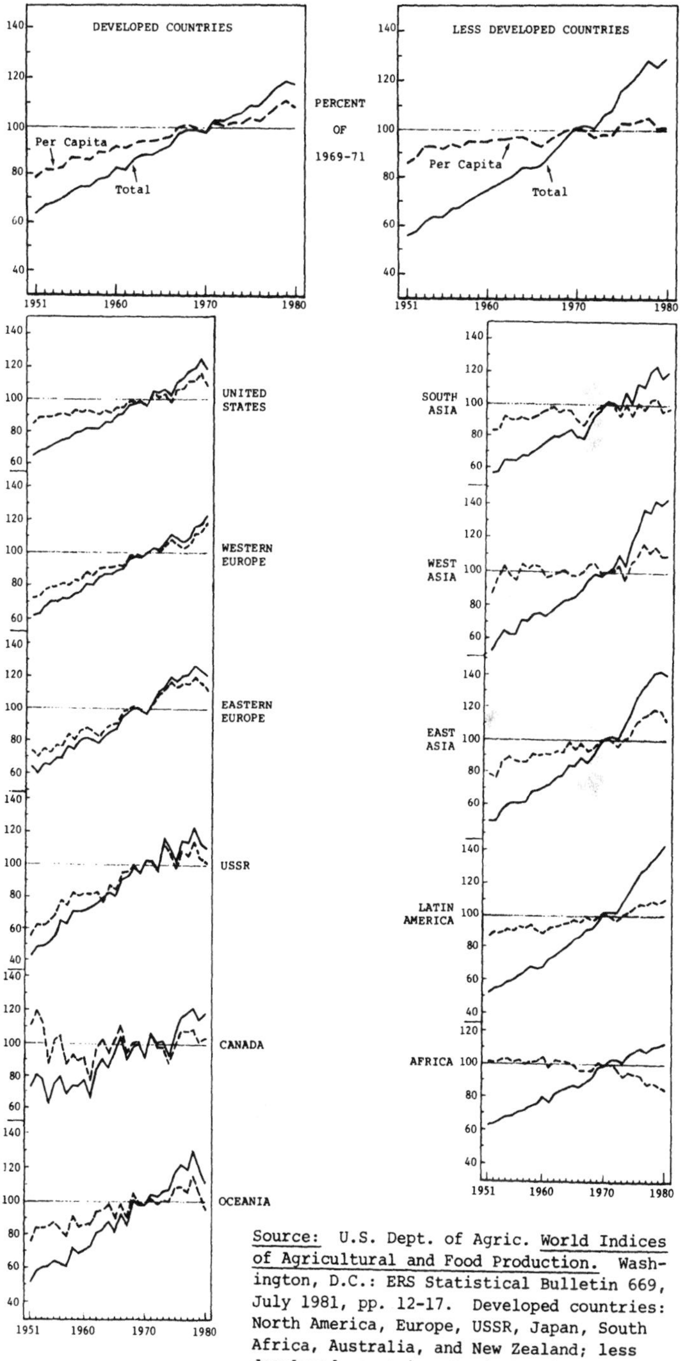

Source: U.S. Dept. of Agric. World Indices
of Agricultural and Food Production. Washington, D.C.: ERS Statistical Bulletin 669,
July 1981, pp. 12-17. Developed countries:
North America, Europe, USSR, Japan, South
Africa, Australia, and New Zealand; less
developed countries: Latin America, Asia
(except Japan and Communist countries),
and Africa (except South Africa).

suggest, in its *State of Food and Agriculture* for 1969, that the food problems of the future might well involve surplus rather than shortage.

The factors underlying the second pause—the "food crisis" of the early 1970s—were more complex and primarily involved the developed rather than the developing countries. In brief, the crisis resulted from an unhappy coincidence of four main influences: an intentional running down of stocks and a holding down of production in the United States; unprecedented prosperity and rising demand in Europe and Japan, which led to rapid increases in the use of grain for livestock feed; a general relaxation of attention to agriculture in the LDCs; and unfavorable weather in India, the African Sahel, and the Soviet Union. The role of the Soviet Union was particularly destabilizing. The failure of its 1972 harvest triggered a run on world supplies and the short crop of 1975 prolonged it. Nonetheless, the crisis was truly global in that the price rises were general and in that it exposed the weaknesses of the international agricultural order. "International" is the operative word: most affected were the countries trading in the world market. Least involved were the largely self-reliant LDC economies.

More recently the pendulum of assessment has swung again. Save in the Soviet Union, harvests almost everywhere were favorable during the latter half of the 1970s. Although some perceive the upturn in the quantity of grain moving from North America to the developing world as a sign that the latter is increasingly unable to feed itself, this interpretation is incorrect. As Figure 3 shows, the bulk of increase is going not as food to the poorest countries, but as feed to those middle-income countries whose populations are becoming sufficiently affluent to effectively demand more meat in their diet. To the extent there is talk among knowledgeable observers of potential crisis, it is usually in the context of the events of the early 1970s and the possibility of a recurrence sometime during the next decade. There have, after all, been few changes in policies that would mitigate against a recurrence following a year or two of below-trend production.

This story of modest progress in the LDCs clearly does not tally with the pessimism of the early FAO and USDA studies. It does not follow, however, that the postwar years have witnessed a reduction in the actual number of people nutritionally distressed. The suggestion that increased production alone could eliminate hunger was only one of the misconceptions conveyed by the early studies.

A second unfortunate legacy was the notion that *countries* could be classified as hungry or well fed. It is now clear that, to the extent that this notion has validity, the early studies misrepresented reality. With food availability estimates that *understated* set against requirement figures that *overstated,* the cards were so stacked that almost all LDCs could

FIGURE 3. WORLD GRAIN TRADE, 1960/61-1980/81

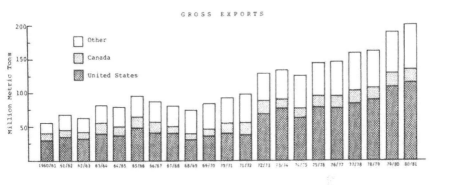

GROSS EXPORTS

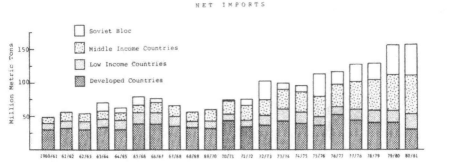

NET IMPORTS

Data from: U.S. Department of Agriculture, Foreign Agricultural Service. Years are July-June marketing years. Low income countries are defined as those with a 1979 per capita GNP of US$670 or less.

be classified as diet deficit. Redone with truly accurate information, it is probable that few countries would be so classified. Much as the protein gap proved a statistical illusion, the list of diet-deficit countries would be whittled away.

But such a computation would perpetuate the most important failing of the early methodology. It is individuals, not countries, who experience nutritional deprivation, and average country data tell us little about the individual. It is now a commonplace among serious pronouncements on the food situation that, equitably distributed, global supplies are sufficient to feed all. The problem is that all within a country do not have equal access to existing supplies, and it is to the impact of this inequality that I now turn.

Measuring the Impact of Income

The first study that attempted to take account of the income effect was FAO's *Third World Food Survey,* published in 1963. Largely the work of the eminent Indian statistician P. V. Sukhatme, then director of FAO's Statistics Division, the study is spotty in its description of methodology. This is understandable. Our insights into the effect of income rest largely on household budget surveys, and if there are few of these today of acceptable quality, there were even fewer two decades ago.

Nonetheless, on the basis of evidence from Maharashtra State in India and elsewhere, Dr. Sukhatme (FAO 1963, p. 51) concluded that "as a very conservative estimate some 20 percent of the people in the underdeveloped areas are undernourished and 60 percent are malnourished. Experience shows that the majority of the undernourished are also malnourished. It is believed therefore . . . some 60 percent of the people in the underdeveloped areas comprising some two-thirds of the world's population suffer from undernutrition or malnutrition or both."

This, of course, was before the recommended allowances for protein were lowered; with revision, the 60 percent malnourished presumably disappeared. On the other hand, Dr. Sukhatme's 20 percent undernourished is not too much different from FAO's current estimate of persons suffering protein-calorie deprivation.

In documentation prepared for discussion at the World Food Conference of November 1974, the FAO took due account of the 1971 reduction in protein allowances and also employed the 1.2 BMR criterion for minimum energy needs for the first time. Though this yielded floor values well below the energy requirement figures used by Dr. Sukhatme—1,500 as opposed to 2,300 calories for the Far East—the proportion of LDC population whose estimated intake fell below it actually increased: from 20 percent to 25 percent.

It is difficult to imagine how this increase came about, and the FAO offered no explanation. My suspicion is that the figures were derived less through research than through a political decision imposed from on high. Few bureaucrats wish to admit that the problem they are relieving is a modest one, and international bureaucrats are no exception. As Table 1 shows, the figures almost certainly contain an element of arbitrariness. Between April 1974, when the preliminary documentation was released, and the conference itself in November, the estimated LDC population with intakes falling below 1.2 BMR was raised from 360 million to 434 million—or from exactly 20 percent to exactly 25 percent.

I confess to similar skepticism about the findings of the two most recent studies that attempted to measure the impact of income—the

TABLE 1. WORLD FOOD CONFERENCE: PRELIMINARY AND FINAL ESTIMATE
OF NUMBER OF PEOPLE THOUGHT TO HAVE HAD AN INSUFFICIENT
PROTEIN-CALORIE SUPPLY IN 1970, BY REGION a/

Region	Population	Percentage Below 1.2 BMR		Number Below 1.2 BMR	
		Preliminary Estimate	Final Estimate	Preliminary Estimate	Final Estimate
	(millions)			(millions)	
Developed	1,074	3	3	28	28
Developing b/	1,751	20	25	360	434
Far East b/	1,020	22	30	221	301
Latin America	283	13	13	37	36
Africa	273	25	25	68	67
Near East	171	20	18	34	30
World b/	2,825	14	16	388	462

Preliminary estimates from United Nations, World Food Conference.
Preliminary Assessment of the World Food Situation, Present and Future.
Rome, April 1974, p. 39. Final estimates from United Nations, World
Food Conference. Assessment of the World Food Situation, Present and
Future. Item 8 of the Provisional Agenda. November 1974, p. 66.

a/ Principal modifications shown in italics.

b/ Excluding Asian centrally planned economies.

World Bank study carried out in 1976 and FAO's *Fourth World Food Survey,* published a year later. Although they used broadly similar techniques for estimating the effect of income (the main difference was that the 1.2 BMR floor criterion was used by FAO; the old recommended dietary allowances by the World Bank) their conclusions differed wildly. As Figure 1 indicates, the FAO concluded that about 450 million people were suffering from protein-calorie malnutrition; the World Bank put the number at almost 1.2 billion. Not knowing what to do about this discrepancy, the recent Presidential Commission on World Hunger mentioned both figures.

Central to the analysis in both studies was the concept of calorie-income elasticity; that is, of the increment in caloric intake associated with an increment in income. The elasticity or elasticities used by the FAO were not stated; the World Bank study postulated a range of from

0.10 to 0.30 for people just meeting their minimal food needs. Although the reasons for its selection were not specified, a calorie-income elasticity of 0.15 was deemed most appropriate, and on the basis of it and some heroic assumptions about income distribution in Asia, Latin America, Africa, and the Middle East, the study concluded that "56 percent of the population in developing countries (some 840 million people) had calorie-deficient diets in excess of 250 calories a day. Another 19 percent (some 290 million people) had deficits of less than 250 calories a day" (Reutlinger and Selowsky 1976, p. 2).

There are a number of reasons for giving minimal credence to the resulting figure of almost 1.2 billion hungry people. The World Bank analysts were apparently unaware of the tendency for food production in developing countries to be underreported, and their use of the old recommended dietary allowances as surrogates for minimal needs was surprising, to say the least. Furthermore, there are serious problems with the concept of calorie-income elasticity. It misleads by suggesting that the relationship between income changes and changes in energy intake is a simple one, reducible to one tidy figure. This is not the case. Figure 4, prepared ten years ago to summarize the effect income has on nutrient intake in Sri Lanka, is suggestive of the real world. The household budget survey on which it is based was then almost unique. It covered almost 10,000 households representative of the entire country, and it was conducted and analyzed with uncommon integrity.

Yet, even with this survey, one can infer little with confidence about the extent of protein-calorie malnutrition in Sri Lanka. The most important dietary adjustment historically associated with rising income is a decline in the importance of the starchy staple foods—read rice in Asia—as sources of energy and a shift to the more expensive, flavorous foods such as meat, fish, and vegetables. In Sri Lanka this tendency is observable among only the four uppermost income classes (20 percent of the population) and then, because of egalitarian measures imposed by the government, only weakly so. Between the lowest class (43 percent of the people) and the next lowest (37 percent), the sole change is quantitative. There is a difference in apparent per capita daily availabilities of 200 calories and 10 grams of protein, but none in diet composition.

What are we to infer from this? Because increased quantity, not quality, was purchased with increased income, the jump from 2,050 to 2,250 calories could be interpreted as implying behavior consistent with enforced reduced activity among the very poor (or actual physical deterioration) and that the 1.2 BMR energy floor of 1,500 calories is an unrealistically low figure for minimum needs in Sri Lanka. But just as reasonably, one might postulate caloric adequacy among the element

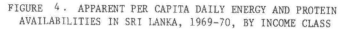

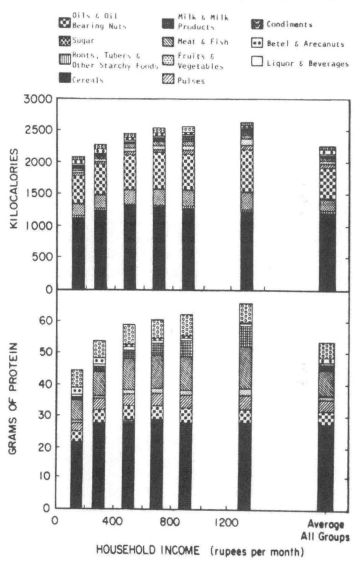

FIGURE 4. APPARENT PER CAPITA DAILY ENERGY AND PROTEIN
AVAILABILITIES IN SRI LANKA, 1969-70, BY INCOME CLASS

Reproduced from: Poleman, Thomas T. Income and Consumption:
Report to the Government of Sri Lanka. Food and Agriculture Organiza-
tion, Nutrition Division and United Nations Development Program Report
No. TA-3198. Rome, 1973. pp. 18-19. Based on data from: Sri Lanka,
Department of Census and Statistics. Socio-Economic Survey of Sri
Lanka, 1969/70, Special Report on Food and Nutrition Levels in Sri
Lanka. 1972.

of society that is too poor to waste anything and which, given the very high rate of unemployment in Sri Lanka, leads a less active life and therefore has lower energy needs. Thus it is possible to have it either way: depending on your assumptions, you can prove beyond a statistical doubt that 43 percent of Ceylonese suffer protein-calorie malnutrition or none do.

Income, then, is clearly crucial, but hasty evaluations of its impact are out of place. If its effect is to be properly understood, the evidence at hand must be subjected to painstaking scrutiny.

An essential backdrop to such scrutiny is an understanding of how poor people over the centuries have contrived to feed themselves and how their diets have changed as they became more wealthy. This was the subject of some truly pioneering research carried out during the 1930s and 1940s by Merrill K. Bennett. He noted that the very poor everywhere seek to maximize the nutritional return per outlay for food by building their diets around foods composed principally of starch: wheat, rice, potatoes, cassava, and the like. This is so because of the cheapness of these starchy staples, whether expressed as market price or production cost. Far less land and far less labor are needed to produce a thousand calories of energy value in the form of the starchy staples than in any other form of foodstuff. Meat producers are inefficient converters by comparison; an animal must be fed between three and ten kilograms of grain for it to produce a kilogram of meat. But most people enjoy meat, and they turn away from the starchy staples as they become wealthier.

A simple way to rank diets is according to the percentage of total calories supplied by the starchy staples and an easy way to record change is to monitor shifts in this starchy staple ratio. In the United States the ratio stood at 55 percent a hundred years ago, when our great-great-grandparents consumed large amounts of bread and potatoes. Today our diets are dominated by meat, fats and oils, sugar, vegetables, and dairy products. We pay more for this diet and presumably enjoy it more, but it does not follow that it is a better diet.

Faced with so much conflicting material about the extent of world hunger, the USDA a few years ago asked me to look into the basis for the confusion. Among my conclusions was one that could have been anticipated at the outset: there still is not enough evidence about the effect of income on food behavior for us to generalize with confidence. Such evidence must come from household budget surveys, and carefully conducted surveys of broadly representative samples are still rare for the developing countries.

About halfway through my research for USDA, it occurred to me that because it was not likely that the next few years would bring more

FIGURE 5. NORTHEAST BRAZIL: APPARENT PER CAPITA DAILY CONSUMPTION OF
MAJOR STARCHY STAPLES AMONG LOW-INCOME CLASSES, 1974-75

(calories)

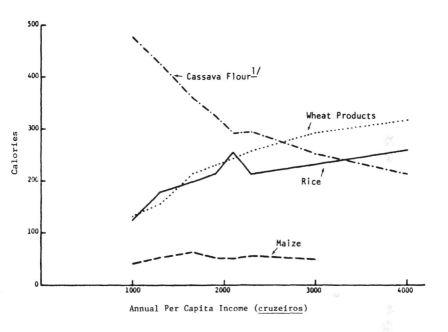

1/ Plus other roots and tubers (5 percent of total).

*Data from: Neville Edirisinghe, "Behavioral Thresholds as Indicators of Perceived Dietary Adequacy or Inadequacy" (unpublished Ph.D. dissertation, 1983).

accurate estimates of either food availabilities or minimal nutritional needs in the developing countries, we should consider abandoning altogether the approach of comparing availabilities with needs. We could search instead for behavior indicative of perceived dietary adequacy or inadequacy. Would not Bennett's progression of dietary change provide the necessary framework, and would not the point where households began to purchase quality instead of quantity be the threshold we sought?

Preliminary analysis by my students of five recent surveys, conducted in Sri Lanka, Indonesia, Bangladesh, Peru, and Brazil, indicates the idea has merit but that it will be of less value in countries where, as in Sri Lanka, one staple is all-pervasive. This is because initial substitution will take place among the starchy staples; and where one food dominates, the substitution will be between quality grades of that commodity, shifts that are extremely difficult to monitor. Figure 5, based on data collected in Brazil in 1974/1975, is suggestive of findings where more than one staple and a clear preference hierarchy prevail. It is apparent that the diet at the lower end of the income range is that of poor people: the four starchy staples (cassava, maize, rice, and wheat) supply over half of the total caloric availability. But it does not appear to be the diet of people threatened with hunger. Additional calories are not purchased as income increases. Instead, consumption of cassava—the least preferred staple—falls off sharply, and its place is taken by rice and wheat bread.

This type of behavior, which is also evident in the data for the other countries examined, is not suggestive of widespread hunger in the LDCs. It indicates instead the ability of the people there to shrewdly allocate their limited resources so as to get by on what, by the standards of the industrialized world, is very little. It is clear that the poor were not excluded from the surveys. The Brazilian data are for the northeastern part of the country, Brazil's poorest region, and the survey has been suppressed by the government because of the social inequality it reveals.

Remedial Approaches

It may appear that, while feeling free to criticize the work of the FAO, USDA, and the World Bank, I have cleverly avoided offering any numbers of my own. Before doing so, I would like to discuss in a general way the various strategies put forth for eliminating hunger. Since the nutritional problems of the developing countries are less a reflection of insufficient production than of the poor's inability to effectively demand a satisfactory diet, these strategies must take into account the extent to which the normal course of development can be expected to raise income.

The Income/Employment Backdrop

Increased levels of food production will, of course, continue to be required, but the evidence for most areas indicates that this increase is forthcoming. I have noted that the record of LDC agriculture over the past quarter century has not been unimpressive; despite the minimal attention given to agriculture in most LDCs, output has expanded no less rapidly than in the industrialized countries and, save in Africa, has more than kept pace with population growth.

What is all the more remarkable is that this has taken place without the huge upsurge in yields that has so transformed agriculture in the developed countries. Figure 6, which compares the production, area, and yield of all grains immediately after the war and recently, offers some perspective. Prior to the war, grain yields everywhere averaged

FIGURE 6. WORLD PRODUCTION, AREA AND YIELD OF GRAINS, AVERAGE
1947-52 AND 1979-80

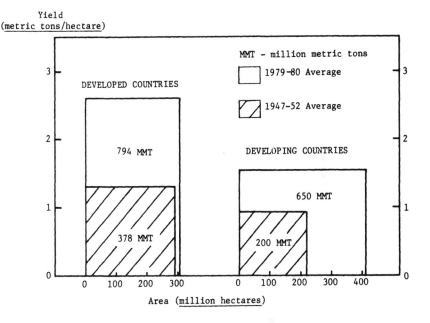

Data from: U.S. Department of Agriculture, Foreign Agricultural Service.

in the neighborhood of one metric ton per hectare. By 1980, as the result of greater use of fertilizer, improved seed varieties, and better cultivation practices, yields had more than doubled in the developed countries. In the LDCs, on the other hand, the gains have been comparatively modest. Despite the publicity given to the Green Revolution and certain spectacular accomplishments, LDC average yields have increased to only a bit above those of the developed countries at the beginning of the postwar upsurge.

The systematic application of the scientific method to food farming in the developing world is very recent, dating back no further than the mid-1940s. It is not surprising, then, that the scope for improving yields has been only superficially exploited. Moreover, breeding work until just this decade ignored the root crops and concentrated on wheat, rice, and maize; even for these crops yields have risen to only a fraction of the potential. A substantial share of the rice produced in Bangladesh and Thailand is of the floating variety, able to grow up to a foot per day if flooding demands it. It is hardly an exaggeration to say that work is only beginning on this unusual crop. The payoff is likely to be great.

If the LDCs are potentially capable of enormous increases in food production, one cannot be equally sanguine about the outlook for providing productive employment for their populations. In part, this pessimism recognizes the selective nature of all economic change—some are caught up in the process, others left out—but above all it mirrors the number of people who will be entering the labor force between now and the year 2000.

The number is truly staggering. Figure 7 illustrates a projection prepared a few years ago by the International Labor Office (ILO). Between 1970 and 2000 the LDC labor force is expected to double— from about 1 to 2 billion people. The billion new jobs that must be found are roughly twice the number presently available in the industrialized countries; this means that the LDCs will need to transform themselves at a rate and on a scale unprecedented in history. In terms of just one country, it means that during the remainder of the century Mexico will add to its labor force each year about the same number of new entrants as the United States and Canada *together* were able to absorb during the boom years of the 1950s and 1960s.

The ILO projects that few of the new entrants will be absorbed into agriculture, and one must ask whether this need be so. The basis is the selectivity of the various technical breakthroughs that have so far characterized the Green Revolution. To the layman the term "Green Revolution" conjures up visions of miracle seeds that offer all farmers the same potential for dramatic increases in yield. In fact, the high-yielding varieties (HYVs) are not designed to be introduced alone, but as one component of a package involving a host of complementary inputs; fertilizers, adequate water, and effective control of disease, insects, and weeds. The miracle rices are highly responsive to fertilizer and yield well only under irrigated conditions. Simply to provide the proper conditions for them to be introduced can be time-consuming and expensive. To the extent that the new systems are specific to particular ecological conditions, benefits will clearly be restricted. Equally obvious is that those best able to afford the new inputs—the larger and wealthier farmers—will reap the lion's share of the benefits.

The experience of Mexico is a case in point. Mexico was the site of the first "agricultural miracle" of the postwar period, thanks to the Rockefeller-funded Office of Special Studies (now CIMMYT—Centro Internacional de Mejoramiento de Maíz y Trigo). The achievements were impressive. The output of maize increased from about 3.5 million metric tons during the late 1940s to 9 million metric tons in 1968. Average yields per hectare almost doubled, going from 700 to 1,300 kilograms. The performance of wheat was even more spectacular: from 300,000 metric tons to over 2.5 million metric tons in just 20 years,

FIGURE 7. ECONOMICALLY ACTIVE POPULATION, RECENT YEARS
AND PROJECTIONS TO 2000

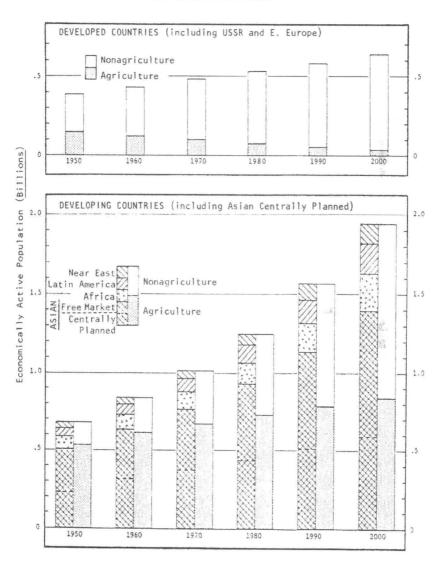

Data from: International Labour Organization. Labor Force Projections. Geneva, 1971. United Nations, Food and Agriculture Organization.
The State of Food and Agriculture--1973. Rome, 1973, pp. 131, 138.

with yields quadrupling to 3.2 metric tons per hectare. Cotton and other crops for export fared almost as well.

But as every Mexican knows, this extraordinary achievement was localized both geographically and with respect to its impact on the rural population. Change has largely been confined to the north and the northwest, where the program of government-sponsored irrigation opened up expanses of highly productive land, and the Gulf coastal strip, which is the one portion of the country receiving abundant rainfall. Though lip service throughout the period was paid to continued agrarian reform, the great majority of the rural population was bypassed. Today less than 5 percent of farm units occupy almost two-thirds of the irrigated area and account for over half the value of production. In contrast, 85 percent of farms have access to only 4 percent of irrigated land and contribute a mere 20 percent of output by value.

The selective impact of their breakthroughs came as a shock to the Rockefeller scientists and their Mexican colleagues, but it shouldn't have. A similar selectivity characterized the innovations which transformed agriculture in Europe and North America during the nineteenth century. But here the historical parallel begins to break down. There is a great difference between the cities of last century's developing countries and those of today's. A hundred years ago the bypassed or displaced farmer could look to the city for opportunity. Industry was growing, and as industry then had high labor requirements, virtually all who left the land found jobs. Today the movement to towns rests on less solid foundations. Though urbanization in the LDCs is proceeding at a breakneck pace, most of the industrialization taking place requires capital and not labor. To a depressing extent, most urban centers continue as administrative and trading centers. Suitable jobs are far fewer than people in search of them.

The prospect, then, is for two groups of persons to be excluded from the course of development: those bypassed by progress in the countryside and the underemployed of the towns. The proportion of the population presently falling within the two groups is anybody's guess. Governments in the developing world do not collect data on unemployment and underemployment, and if they did the findings would be too politically distasteful to permit release. The World Bank and other international agencies speak of the "lower 40 percent," and even if this figure is a very rough estimate, it is probably a fair one. Somewhere between one-quarter and one-half of the population of the developing world is being bypassed by the forward march of economic change.

Recognition that the income effect cannot be relied on to eliminate all nutritional deficiencies within the lifetime of many of those most seriously afflicted has given rise to a host of alternative proposals. Their number, in the words of one observer (Chafkin 1978, p. 7), is indicative "of how much of a growth industry human nutrition and world hunger activity has become." The proposals can be divided into two broad groupings: those that focus on correcting the causes of economic inequality, and those that aim at affording a measure of short-term relief.

Proposals in the first category range from the hopelessly naive to the outright revolutionary, but include a broad center setting forth conditions for "growth with equity." These include:

- An emphasis on developing the rural sector and on the immediate implementation of land reform.
- The retargeting of production toward meeting the minimum needs of all.
- The use of "appropriate"—by which is meant labor-intensive—technology.
- A willingness to view the Chinese and Cuban experiences with tolerance.

Proposals of this type have taken on a certain air of dogma among nutrition planners at the World Bank and the United States Agency for International Development (USAID), but one wonders to what end. Since so many of the changes visualized fly in the face of the interests of established elites, they are not likely to be implemented without protracted struggle. Furthermore, there is the very real danger—as El Salvador demonstrates—that any such struggle would come to be seen in terms of the East-West conflict.

Treating the Symptoms

Given these political problems and the magnitude of the employment dilemma, my suspicion is that both the West and a majority of LDCs will not attempt to resolve the causes of poverty in a generation, but will seek instead to treat the symptoms. Can those excluded from the development process be helped to achieve a minimally acceptable diet? The answer is yes, but the task will not be easy.

Measures for easing the plight of the nutritionally deprived without transforming the social structure include those designed to increase food consumption without a corresponding rise in food expenditures and those that improve the nutritive value of given foods.

Fortification of traditional foodstuffs with special nutrients is the most attractive means of accomplishing the latter. Iodization of salt is a classic example, as is the admixture to milled cereals of niacin, iron, thiamine, riboflavin, and calcium. The impact of such measures can be widespread and immediate. Beriberi, for instance, was endemic on the famed Bataan Peninsula of the Philippines immediately after the second world war. With the introduction of thiamine enrichment of rice, it virtually disappeared. The problem with this approach, of course, is that enrichment is only possible when foods are centrally processed; and in the very poor countries few are.

Similarly flawed are most of the schemes designed to permit greater consumption of food for a given level of expenditure. These usually involve some form of price manipulation by government, whether through direct procurement in the countryside, by subsidizing aspects of production or consumption, or by controlling the price paid by consumers.

Virtually all developing countries have one or more programs of this type, and as their effect is to transfer income, the motivation has been more political than nutritional. Nonetheless, their nutritional impact can be appreciable. In Sri Lanka, for instance, rice was "rationed" from the second world war to a year or two ago; virtually the entire population was entitled to a weekly quantity either free or at a subsidized price. Throughout most of the period this amounted to two pounds (0.91 kgs) per head per week, or the equivalent of about 475 calories per day. Most observers credit this program with having contributed to the well-being of Sri Lankan people. The same is true of the Egyptian program of subsidizing the retail prices of staple foods in Cairo and Alexandria. However, the cost of such schemes can be high. The rationing program in Sri Lanka regularly absorbed between 15 and 20 percent of the government's budget, and was abandoned by the current leadership as being incompatible with rapid economic growth. One estimate of the cost of the Egyptian food subsidy put it at one-twelfth of the country's GNP. Neither program would have been possible without concessional food from abroad.

In addition to the expense, the problems with programs of this type are several and severe:

- They tend to be restricted to the urban centers—the Egyptian and Bangladesh cases being the most prominent—and thus have minimal impact on the very poor, most of whom still live in the countryside. In Bangladesh, proof of employment is a prerequisite for being issued a ration card.

- If not political from their inception, they quickly become politicized, with the result that modification is difficult. An attempt to reduce the consumer food subsidy triggered such severe riots in Cairo in January 1977 that the plan was immediately rescinded. The price and wage distortions they engender thus tend to become permanent.

- Where not dependent on food aid from abroad, the schemes tend to rely on low procurement prices to keep costs down. In either case the effect is to discourage the growth of domestic agriculture.

This disincentive effect is probably the most telling objection to both food aid and the various distribution and subsidy schemes as vehicles for combating malnutrition. To get around it a number of two-price systems and other devices have been proposed. By far the simplest and—when it is remembered that hunger is not all-pervasive—the most appealing of these is to channel assistance in kind directly to those at greatest risk through maternal and child health clinics.

To some, this suggestion no doubt will seem a step in the wrong direction. For it is precisely this type of targeted intervention that was long central to the (modest) activities of the more traditional nutritionists working in developing countries and that has been downgraded by the new wave of nutrition planners in the World Bank, FAO and USAID. Yet there are, I believe, compelling reasons for its use.

Assisting the Nutritionally Vulnerable

If we do not know how many among the poor in the LDCs suffer nutritional deprivation, nutritionists are agreed that the preschool child and the pregnant and lactating mother are those most likely to be adversely affected by protein-calorie malnutrition (PCM). There are several reasons for this. The early growth and reproduction phases are nutritionally the most demanding in the life cycle:

- The total energy cost of a pregnancy is estimated to average between 40,000 and 80,000 calories.

- To produce 850 milliliters of milk daily during lactation, a mother's energy needs will be about 600 calories above normal.

- Because they are growing so rapidly, the infant and young child require more than twice the number of calories per unit of body weight as do adults and half again as much as adolescents.

Yet it is precisely the mother and young child whose needs are often least reflected in the choice of foods purchased by the household and who may be the residual claimants on what has been prepared for all to eat.

Discrimination against mothers and the young in feeding habits reflects educational as well as income deficiencies. It is not just that undesirable food taboos relate particularly to the mother and her young;

where households do not eat together, the father and elders will typically satisfy themselves first, and the women and children get what is left. Adult tastes, rather than those of the infant, will be the usual criteria of dietary excellence, with the result that much of the animal protein a meal contains can be impossible for the very young to swallow. The younger the child, moreover, the less well he is able to fend for himself at the table. In times of shortage the mother is likely to defer to her children, not realizing that it is not only she but also her unborn child who will suffer.

In its extreme form, PCM among the young occurs either as kwashiorkor or marasmus or sometimes as a combination of the two. Kwashiorkor is generally the result of an inadequate intake of protein relative to calories. It typically affects children who, after a period of breast feeding, are weaned onto starchy staples low in protein—such as the tropical roots and tubers—or sugary foods. It is most common in those parts of tropical Africa where roots, tubers, and bananas are the dominant starchy staples; indeed, kwashiorkor comes from the Ga and means "disease that occurs when displaced from the breast by another child."

Marasmus, on the other hand, arises from an insufficiency of both energy intake and protein. The condition usually occurs in the first year of life and among the children of undernourished women. They are commonly of low birth weight and even in their first few months will show large weight-for-age deficits. Marasmus is by definition chronic, whereas kwashiorkor is an acute condition amenable to rapid reversal.

Cases of pure kwashiorkor or marasmus seem to be the exception rather than the rule. Instead, most severely malnourished children will present signs and symptoms of both conditions and perhaps even alternate between the two. In addition to low weight for age and other overt physical signs, symptoms include apathy, instability, and poorly developed motor skills.

Although marasmic children may be at risk of outright starvation, the main danger to the severely malnourished lies in their diminished resistance to other disease. Should they survive, they may well go through life permanently impaired, both mentally and in the height and weight they will ultimately attain. Although the linkages between severe PCM and brain growth and development are by no means understood, they give rise to particular concern.

Compared with its impact on pregnant and lactating mothers and the very young, the adverse effects of PCM on the other elements of a population are likely to be moderate. (Exceptions, of course, are the aged and the infirm; on them the impact can be devastating.) This is because these people are either not growing so rapidly or have stopped growing altogether and can adapt to reduced energy intake by either taking off body weight or by curtailing activity.

The human body is remarkably adaptive to reduced levels of food intake. Controlled studies among adults have shown that if caloric intake is cut to 50 percent of normal, body weight will drop within a few months by about a quarter. Thereafter a reduced level of activity can be maintained for many months before additional weight loss sets in and the incidence of nutrition-related disease rises. Thus the aftereffects among adults of war-induced privation in the Netherlands in 1944–1945 and in Biafra in 1968–1969 are not thought to have been lasting; certain groups in Africa may experience with no apparent impairment significant changes in body weight due to the preharvest hunger phenomenon. This is not to suggest that the lethargy often observed among the poor in developing countries is not an adjustment to inadequate diet; only that it is reversible and need only be temporary.

Some nutritionists have suggested that cases of frank marasmus and kwashiorkor among the very young are but the tip of an enormous iceberg of PCM in the developing world and that for every child demonstrating symptoms of clinical PCM there may be 99 others who are inadequately nourished, grow poorly, and are highly susceptible to disease infection. The iceberg analogy may indeed be appropriate, but should be treated with skepticism. Certainly attempts to quantify the submerged portion—where no demonstrable harm is being done—do not yet warrant scientific credence.

Estimating the extent of PCM among young children involves the incipient science of nutritional anthropometry, and debate attends the standards for healthy children it should employ, the measurements it should involve—whether weight for age, height for age, or weight for height—and where the cut-off criteria should be established. The reference standards usually employed are those of the U.S. National Academy of Sciences and are defended on the grounds that within the 6- to 71-month age group possible distortions resulting from individual or racial differences will be minimal.

Very few follow-up studies of the health experience of children displaying specific anthropometric deficiencies have been carried out, and until such evidence is in hand and country-specific benchmarks have been determined, anthropometric evaluations should be interpreted with caution. Nevertheless, it has become common practice to define severe PCM as being evidenced by a weight-for-age of less than 60 percent of standard and the moderate form as being reflected by weights in the 60–80 percent range. Following (more or less) these definitions, the findings of the hundred or so anthropometric surveys available a decade ago have been summarized in Figure 8. It is apparent that only

FIGURE 8.   PREVALENCE OF SEVERE AND MODERATE PROTEIN-CALORIE
MALNUTRITION, BY REGION, 1963-73

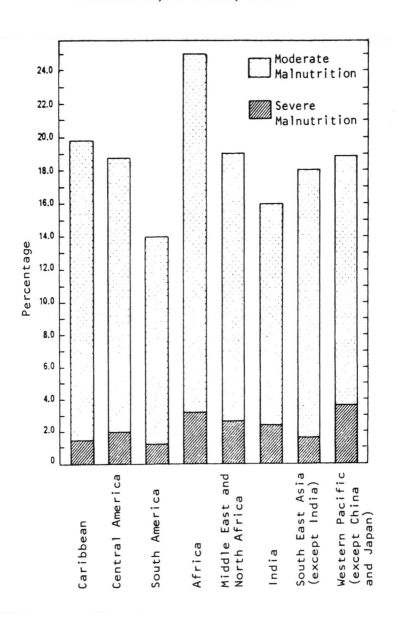

Reproduced from Bengoa, J. M., and G. Donoso. "Prevalence
of Protein-Calorie Malnutrition, 1963 to 1973." PAG Bulletin, vol. IV, no. 1 (1974), p. 31.

a small fraction of children—in most of the surveys just 1 or 2 percent—suffered from severe PCM, and perhaps for each severely afflicted child, 10 or 15 were moderately afflicted. Beyond this, generalization is impossible. Most of the surveys covered small samples of questionable typicality, and the range in findings is too extreme to permit extrapolation. If the number of those at greatest risk is to be estimated, therefore, we have no alternative but to do so rather arbitrarily.

Table 2 offers one possible method. The number of pregnant and lactating women is approximated by doubling the birth rate. To this figure is added the number of infants below five years of age. Arbitrary percentage estimates of those likely to be at risk nutritionally are then applied. A 10 percent assumption is not an unreasonable minimum for most developing countries, while a 50 percent figure would seem an absolute maximum. The resulting range of those at risk—62 million to 309 million persons—defines a world hunger problem of vastly different dimensions than that conjured up by the World Bank study and the *Fourth World Food Survey* (Table 3). Although Table 2 includes China, which the two studies did not, even the 50 percent assumption yields figures well below the number of those considered to have an insufficient protein-calorie supply in the *Fourth World Food Survey* and about one-quarter the number arrived at by the World Bank study.

About a year ago I was quoted in a front-page article in the *New York Times* (5 October 1981) to the effect that "food aid of only three million tons of grain a year, if it could be channeled to the truly needy, would enable 100 million malnourished people to have an adequate diet." Whence came this statistic? It was a guess, worked out during a telephone interview, that perhaps 100 million mothers and young children are seriously malnourished and their average deficiency is in the neighborhood of 300 calories per day. I cannot defend it. The truth could be 400 calories for 150 million people or 200 calories for 200 million. If the former, the deficit would be 6 million metric tons; if the latter, 4 million tons. The point is that the deficits are trivial for a world that produces about 1.5 billion tons of grain and in which about 200 million tons move annually in international trade. There is no doubt in my mind that such quantities would be immediately forthcoming—as FAO's unexplained target of 10 million tons of food aid has not been—if the donors could be assured that the recipient countries would pass them on to those in true need.

Herein lies the rub. Few developing countries have so far been able to muster the medical expertise and adminstrative competence to organize a nationwide system of maternal and child health clinics. In Sri Lanka, for instance, it is reckoned that at best perhaps only two-thirds of the children needing it are benefited by the country's Thriposha weaning

TABLE 2. NUMBER OF WOMEN AND CHILDREN AT RISK NUTRITIONALLY
ACCORDING TO TWO ASSUMPTIONS, 1975, BY REGION

(millions)

Region [a]	Total Population	Infants (Age 0-4)	Births	At Risk [b]	
				10 Percent Assumption	50 Percent Assumption
Far East (ex-China)	1,057	174.3	43.5	26.1	130.6
China	823	98.6	22.1	14.3	71.4
Africa	331	60.6	15.8	9.2	46.1
Latin America	319	50.4	11.7	7.4	36.9
Near East	188	31.9	8.1	4.8	24.0
TOTAL	2,718	415.8	101.2	61.8	309.0

Population data from: World Bank. World Atlas of the Child. Washington, D.C., 1979.

a/ Regional breakdown follows current FAO usage and includes only those countries classified as "developing." See: United Nations, Food and Agriculture Organization. The Fourth
World Food Survey. Statistical Series 11, 1977.

b/ Assumptions are that 10 percent and 50 percent of the vulnerable groups (infants up
to five years of age and pregnant and lactating mothers) can be considered malnourished. The
number of pregnant and lactating mothers is taken as being twice the birth rate.

TABLE 3

FAO and World Bank Estimates (in Millions) of Persons Suffering PCM Compared with Author's Assumptions of Women and Children at Risk, 1965-1975

	Insufficient Protein-Calorie Supply		Vulnerable People at Risk	
	Fourth Survey	World Bank	10 Percent Assumption	50 Percent Assumption
Far East (ex-China)	297	736	26	131
China	--	--	14	71
Africa	83	190	9	46
Latin America	46	113	7	37
Near East	29	91	5	24
	455	1,130	62	309

Sources: FAO Fourth World Food Survey. (Statistics Series No. 11, Rome, 1977), p. 53. Shlomo Reutlinger and Marcelo Selowsky. Malnutrition and Poverty. Washington, D.C.: World Bank Staff Occasional Paper 23, 1976, pp. 24, 53-55.

food program; and Sri Lanka is recognized to have one of the best public health systems in the developing world.

Yet establishing such public health systems should be a priority matter everywhere, if only to help bring the birth rate under control. It is now clear that rapid population growth can be contained rather quickly once certain preconditions have been achieved. Among the most important of these preconditions is a reduction in infant mortality, so that parents need no longer plan on two live births in order to feel reasonably assured that one child will reach maturity. To this end there are no more effective means than systems providing supplemental food as well as medical services to mother and child.

Happily, there is a growing body of evidence that suggests these public health systems need not be as sophisticated as was once thought. This is certainly a lesson of the Chinese "barefoot doctor" program. If life expectancy in China is indeed half again what it was in 1949, the potential of relatively simple paramedical services should not be underestimated.

Similarly, new simplified methods of treatment of disease would seem to hold great promise. A case of point is the use of oral rehydration therapy in the treatment of diarrhea. Diarrhea and its consequences are among the major causes of infant mortality, with perhaps 500 million episodes annually among preschool children worldwide.

Since the immediate causes of death from diarrhea are a loss of water and electrolytes, the essence of therapy is the replacement of these substances. The traditional method has been intravenous—a form of treatment that requires the child be brought to a hospital or health care center. In the past few years, however, it has been found that oral administration of a mixture of water, electrolytes, and glucose in the home can have the same beneficial effects. A recent trial of the technique in rural India resulted in a reduction in fatalities to one-fifth their previous level. There are no doubt other innovations whose impact could be equally dramatic.

The Special Problem of Africa

I indicated in the introduction that I felt the hunger problem was now manageable and could be eliminated in virtually all parts of the world before the year 2000 given proper governmental resolve. That this will come to pass in Latin America and Asia I am reasonably certain. Whether it will also happen in sub-Saharan Africa is very much in doubt.

It is customary to think of hunger as being primarily a problem of Asia, and so it once probably was. But today the incidence of serious malnutrition is highest in Africa. Black Africa is the only region of the developing world where, if the statistics are to be believed, per capita food production has not risen, but declined during the past two decades. It is also the only region where life expectancy is still under 50 years and where the birth rate has not begun to decline. And finally (if we exclude Cambodia) it is the only region to have experienced outright famine since the second world war.

Why this dismal performance? It certainly does not stem from a shortage of resources. The amount of arable land per cultivator in Africa is more than double that in Asia. And the infrastructure left by the departing colonial administrators was not inappropriate for sustained economic growth.

The answer, of course, lies in political instability and the incompetence and corruption that have characterized so many African governments since independence.

The two countries I know best—Ghana and Uganda—are depressing cases in point. I spent some months in the early 1960s studying the system through which Ghana's cities were fed, and was mightily impressed with its efficiency. Commodities moved smoothly into Accra from up to 300 miles away and there was rational substitution in the availability of staple foods from one season to the next. At independence Ghana was the wealthiest country in tropical Africa. Today, 20 years and five coups later, it is no longer. The cause is human folly. Ghana's once thriving cocoa industry is producing at a level 50 percent below that of the 1960s—not because there is no demand for the product, but because the government, in an effort to enhance its revenues, compels growers to accept a price rather less than half that prevailing on the world market. A chicken sells in Accra today for the equivalent of about one-tenth of a senior civil servant's monthly salary; fish, the poor man's protein, sells for about three days labor per pound—not because the market women have lost their enterprise, but because fuel has become so expensive and vehicles and roads have so deteriorated that they can no longer exercise it.

The Uganda story is even sadder. Makerere University, where my family and I spent a delightful year in the late 1960s, had an excellent academic tradition and was home to first-rate medical and agricultural faculties. While studying the marketing system I was again impressed with the liveliness and efficiency of producers and traders, and the wealth of the country. I suppose the less that is said of conditions today the better. Famine apparently is common among the Karamojong in the northeast, with armed helicopters needed last year to ride shotgun on relief shipments. The gentleman with whom I studied Kampala's food supply subsequently rose to become dean of Makerere's Faculty of Agriculture and pro-vice-chancellor of the University. Subsequent to that he had to disguise himself as a peasant and flee for his life from Idi Amin's goon squads. Makerere is in a shambles.

What happened in Uganda is an extreme instance of the collapse of law and order in Africa. But it affords little satisfaction that the Ghanaian experience is more typical. A recent World Bank review concluded that just about every African country enforced, for the same political reasons as in Ghana, pricing policies that discourage the production of food. Once in place, such policies are extremely difficult to reverse.

The history of hunger and famine is not easy to trace. But to the extent that it is, it is evident that causality has laid more with manmade than with natural events. If the twenty-first century witnesses the persistence of hunger in Africa or elsewhere, we will only have ourselves to blame.

References

I have for many years been interested in questions relating to world hunger and have, with my students, written extensively on the subject. The ideas summarized in this paper have been drawn freely from the following overviews, in which may be found more detailed citations to the literature than given here:

Poleman, Thomas T. "Alimentos-población-empleo: Implicaciones en el desarrollo de México." In *Alimentación: Reto de México,* ed. A. Palma-Gómez, pp. 10–22. Mexico City: Fomento Cultural Banamex, June 1979.

_____ . "Employment, Population, and Food: The New Hierarchy of Development Problems." *Food Research Institute Studies* 11, 1(1972): 11–26.

_____ . "Famine." In *Academic American Encyclopedia* 8:18–19. Princeton: Arete Publishing Company, 1980.

_____ . "Food and Population in Historical Perspective." In *Food, Population, and Employment,* ed. T. T. Poleman and D. K. Freebairn, pp. 3–18. New York: Praeger Publishers, 1973.

_____ . "Food, Population, and Employment: Ceylon's Crisis in Global Perspective." *MARGA* 1, 3(1972):25–50.

_____ . *The Thomas Robert Malthus Lecture—The Food-Population Prospect: 175 Years Later.* Institute of Nutrition Occasional Paper, Vol. 1, No. 10. University of North Carolina, Chapel Hill, July 1977.

_____ . *Income and Consumption: Report to the Government of Sri Lanka.* Food and Agriculture Organization, Nutrition Division and United Nations Development Program Report No. TA-3198. Rome: FAO, 1973.

_____ . "Quantifying the Nutrition Situation in Developing Countries." *Food Research Institute Studies* 18, 1(1981):1–58.

_____ . A Reappraisal of the Extent of World Hunger." *Food Policy* 6, 4 (November 1981):236–252.

_____ . "World Food: A Perspective." *Science* 188, 4188 (9 May 1975): 510–518.

_____ . "World Food: Myth and Reality." *World Development* 5, 5–7 (May–July 1977):383–394.

My work, in turn, builds on that of my mentors a generation ago at Stanford's Food Research Institute, most notably Merrill K. Bennett and Helen C. Farnsworth. Although now somewhat dated, Bennett's *The World's Food* (New York: Harper and Brothers, 1954) remains the classic analysis of the linkages between income and diet, as does Farnsworth's "Defects, Uses, and Abuses of National Food Supply and Consumption Data," *Food Research Institute Studies* 2, 3 (1961):179–201, on our inability to adequately quantify the food economies of developing countries.

Of the various official attempts to measure world hunger, the FAO's series of World Food Surveys is the best. They have improved with time and provide the student with a basis for understanding the evolving conventional wisdom. A fifth survey is scheduled for 1984.

Food and Agriculture Organization. *World Food Survey.* Washington, D.C.: FAO, 1946.

_____ . *Second World Food Survey.* Rome: FAO, 1952.

_____ . *Third World Food Survey.* Freedom From Hunger Campaign (FFHC) Basic Study No. 11. Rome: FAO, 1963.

_____ . *Fourth World Food Survey.* Statistics Series No. 11. Rome: FAO, 1977.

The World Bank study and the USDA studies mentioned in the text are:

Reutlinger, S., and M. Selowsky. *Malnutrition and Poverty.* Staff Occasional Paper No. 23. Washington, D.C.: World Bank, 1976.

U.S. Department of Agriculture. *The World Food Budget, 1962 and 1966.* For. Agr. Econ. Rep. No. 4. Washington, D.C.: ERS, 1961.

_____ . *The World Food Budget, 1970.* Washington, D.C.: ERS, For. Agr. Econ. Rep. No. 19. Washington, D.C.: ERS, 1964.

Other sources I have referred to here are:

Boyd-Orr, Lord John. "The Food Problem." *Scientific American* 128 (August 1950):11–15.

Chafkin, Sol. "The Emerging Concern for Human Nutrition and World Hunger." *Agricultural-Food Policy Review,* AFPR-2. Washington, D.C.: USDA, ESCS, September 1978.

FAO. *The State of Food and Agriculture, 1969.* Rome, 1970.

Paddock, William, and Paul Paddock. *Famine, 1975!* Boston: Little, Brown & Co., 1967.

Presidential Commission on World Hunger. *Overcoming World Hunger: The Challenge Ahead.* Washington, D.C.: GPO, March 1980.

United Nations. *Assessment of the World Food Situation, Present and Future.* Item 8 of the Provisional Agenda, World Food Conference, Rome, November 1974.

3
Hunger: Defining It, Estimating Its Global Incidence, and Alleviating It

T. N. Srinivasan
Department of Economics
Yale University

Introduction

A definition of hunger is that it represents the "complex of unpleasant sensations felt after deprivation of food, which impels an animal or a man to seek food and which is immediately relieved by the ingestion of food." (FAO 1982, p. 20). Since occasional episodes of hunger arising out of temporary deprivation of food are not of policy concern, the Presidential Commission on World Hunger concluded that "the true hunger problem of our time is chronic undernutrition—the problem of millions of men, women and children who do not get enough to eat" (Presidential Commission 1980, p. 20). More specifically, undernutrition is "a pathological state arising from an intake of an inadequate amount of food, and hence of calories, over a considerable period of time, with reduced body-weight as its principal manifestation" (FAO 1982, p. 20). Implicit in this definition is the notion that the energy intake (the energy content of food intake measured in kilocalories per day) required to maintain body weight of an individual while (s)he continues to perform whatever energy-using activity to which s(he) is accustomed is well defined. It is further implied that the major cause of undernutrition is inadequate intake of calories. I will argue in some detail that a biological basis for defining a *fixed energy requirement* for humans does not exist. Nor is the evidence for attributing undernutrition mainly to inadequate energy intake beyond doubt. Estimates of global undernutrition such as those put out by the World Bank (Reutlinger and Selowsky 1976) and the FAO (1977) are of dubious validity.

Naive comparisons of average energy requirements and average intakes of subgroups of populations (as for instance, income or expenditure classes, rural and urban population, etc.) such as those made in the World Bank study should rightly be discarded as meaningless. This does not mean, however, that the proportion of the population of the world (particularly the developing world) that does not get adequate food is negligible, adequacy being defined pragmatically rather than in terms of some pseudoscientific norm for energy intakes. Some producers of food can clearly meet their food needs in large part, if not wholly, but nonproducers of food (as well as small farmers) require access to food through exchange with food producers or through unrequited transfers from them. It is therefore of interest to look both at the distribution of aggregate food output and the access to the available food (through trade and transfer) among countries, as well as among socioeconomic groups within countries. In addition, temporal fluctuations (around the trend) in output will be examined from the point of view of distinguishing periodic supply shortfalls due to unfavorable weather and environmental factors (relative to an otherwise satisfactory trend) from an inadequate trend level of output. This is the issue of food security for countries and for socioeconomic groups within countries.

The latter part of this paper will set the problem of hunger and inadequate and insecure access to food in the broader context of poverty and underdevelopment. In particular, access to agricultural resources such as land, water, technology, and credit, and to opportunities for gainful employment in agricultural and nonagricultural activities will be examined as determinants of the extent of poverty. The paper will conclude with some observations from a policy perspective.

The problem of hunger in the sense the word is used here is of significant and serious proportions mostly in the developing countries, though it is not completely absent in rich countries such as the United States. The focus of discussion will naturally be on the developing countries and such developed countries as are significant traders in the international food market. The second largest developing country in the world in terms of population is India. Being an open society and relatively more advanced among the developing countries, its problems, policies and performance have been extensively documented. Indian data and experience will be extensively used in the following discussion.

Energy Requirements and Undernutrition

One of the widely used methods of indirectly estimating the extent of undernutrition[1] in a population (or subgroup thereof) is to equate it to that proportion of individuals whose daily energy intake falls below some norms or recommended intakes called "requirements." These requirements, for a reference population of reference men and women, are put together and revised periodically by a group of experts appointed by the FAO and the World Health Organization (WHO).[2] The expert groups explicitly warn that "The figures for recommended intakes may be compared with actual consumption figures determined by food-consumption surveys. Such comparisons, though always useful, *cannot in themselves justify statements that undernutrition, malnutrition or overnutrition is present in a community or group, as such conclusions must always be supported by clinical or bio-chemical evidence.* The recommended intakes are not an adequate yardstick for assessing health because . . . each figure represents an *average* requirement augmented by a factor that takes into account interindividual variability" (FAO 1974, p. 2, emphasis added). Furthermore, it is admitted that the methodological basis for estimating energy requirements is weak, and much of the information on protein requirements and energy-protein relationships comes from studies on healthy young men in the United States and is of questionable applicability for other populations (FAO 1978a, pp. 2–4).

In spite of such unambiguous warnings, even the FAO has not refrained from using energy norms (set considerably below the recommended intakes, so as to cover just the energy needs for maintenance bodily functions and not for any activity) for estimating the extent of global undernutrition (FAO 1977). A related misuse of the energy requirements is in determining the extent of poverty in a population by defining a person as poor if (s)he cannot afford to buy enough food to meet the energy requirements, and in using the observed relationship between income and energy intake to determine the poverty level of income or poverty line (Ojha 1970; Dandekar and Rath 1971). The study of Reutlinger and Selowsky (1976) also uses the income-energy intake relationship, except that population groups identified are termed "mal-nourished" instead of "poor." The wild estimates of the global extent of undernutrition thrown up by this methodology, far from identifying and focusing attention on the real and manageable task of eliminating food inadequacy, perhaps induce a sense of despair about the task. It will be argued below that such estimates (not reproduced here) are devoid of significance.

There are several conceptual and measurement problems associated with the use of the recommended average. Even if the energy requirements of a randomly chosen man from a population of reference men is a fixed number, since the published norms are the *average* of the norms of the individuals in the population, it is obvious that half of the population of healthy reference men will have intakes (which equal their requirements, since they are healthy) below the average.[3] The methodology of the studies mentioned earlier goes even further: it ignores interindividual variations in intakes as well as requirements in a group by comparing average intakes with norms.[4] To term a whole population of individuals as undernourished if their average intake is below the average norm is clearly erroneous. This is like determining whether a group of soldiers who cannot swim could cross a river safely by comparing their average height with the average depth of the river!

It is becoming increasingly evident that the hypothesis that adults of the same age, sex, and body composition with similar activity patterns and environments have essentially the same energy requirements is not valid. As such, it is not meaningful to treat the habitual intake of a healthy adult engaged in the same or similar activity from day to day while maintaining body weight as that individual's requirement. A study by Widdowson (1947) found that there were substantial variations (a coefficient of variation of over 15 percent) in the intakes of individuals of the same age, sex, and similar activity pattern, even after adjustments for variations in body weight. In another study by Edholm et al. (1970), energy intakes as well as energy expenditures of army recruits were measured carefully. They found that intakes of the same individual varied substantially from day to day while energy expenditures varied much less. Even more interestingly, the variation in intakes was less than what would be expected if intakes on different days were to be independently and identically distributed. A similar pattern in the variance in intakes was seen in a longitudinal study by Acheson et al. (1980) of body weight and body fat changes in twelve adult males spending a year on an antarctic base.

Sukhatme and Margen (1982), from whose work the above discussion is drawn, conclude that the energy balance (i.e., intake minus expenditure) of a healthy individual maintaining body weight and performing the same activity is a stationary stochastic process with mean zero. In other words, the energy balance on successive days is not independent but correlated. Put another way, humans possess a physiological regulatory mechanism for controlling appetite and energy expenditure so that variations in intakes within certain homeostatic limits are "absorbed" without leading to under or overnutrition. This in turn means that the energy "requirement" is not a fixed number. Bennett and Gurin (1982) describe other experimental evidence to show that energy balance is regulated by internal control systems rather than by conscious management of food intakes. The upshot is that defining undernutrition as a function of food intakes dropping below a fixed energy requirement is fundamentally wrong, and so are all estimates of the extent of global hunger and undernutrition based on this fallacious definition. According to Sukhatme, making allowances for intra- and interindividual variations in intakes would considerably reduce the estimated extent of under-nutrition.

The autoregulatory process described above controls short-term variations in intakes, but there may be a long-term genetic adaptation through changes in metabolic efficiency. Edmundson (1979) concludes on the basis of his study of individual variations in basal metabolic rate (BMR) and mechanical work efficiency in eastern Java that "low energy intakes in East Java may be related to both long-term genetic adaption and short-term phenotypic adaptation with a decrease in BMR playing a major role in facilitating a higher level of metabolic efficiency for energy stress subjects. Data clearly indicate that low levels of energy intake do not necessarily result in low levels of human performance (p. 189). Apfelbaum et al. (1971) had reached a similar conclusion earlier: "The change in energy expenditure constitutes an adaptation phenom-enon—an environmental change (intake) provokes a reaction (change in expenditure), which maintains the organism as constant as possible (energy stores). The fact that this adaptation is progressive shows that it must be realized through an integrating system such as energy stores and also, as the number of adipose cells does not vary in adults, the adipose cell volume. How this information is transmitted or by which mechanism the thermogenic yield is modified by metabolic pathways is not known."

Ferro-Luzzi et al. (1975 p. 1408) analyzed the dietary intakes of 482 New Guinean children aged 1–18 years. Data on body weight and skinfold thickness were also recorded. If the FAO/WHO standards of energy-protein malnutrition and Western standards of growth were applied to their data, "growth retardation and the frequent presence of clinical malnutrition would be anticipated. This was only partly sub-stantiated by an assessment of nutritional status: there was a low prevalence of clinical signs of specific nutritional deficiencies. Only occasional cases of Kwashiorkor and Marasmus were observed in the villages and none were encountered in the survey population. Their incidence and severity did not match the degree and frequency of calculated dietary inadequacy. All cases could be traced back to irregular or prolonged untreated infectious diseases." The authors emphasize "that the high proportion of nutritionally inadequate diets, assessed on an age basis using materially favored Caucasian populations as standards, does not match with physiological or clinical signs and symptoms of malnutrition in these New Guinean children" (p. 1452).

Nick Eberstadt (1981) in his stimulating paper quotes several studies that corroborate the conclusion of Ferro Luzzi et al. A study of USAID showed that 42 percent of Sri Lankan children were moderately or severely malnourished and less than 10 percent were normal when U.S. height-weight charts were used to assess growth. Yet life expectancy at birth in Sri Lanka is over 65 years. The Pan American Health Organization apparently found that nearly half the population of Barbados, Costa Rica, Guyana, Jamaica, and Panama suffered from some degree of malnutrition. Yet in each of these countries life expectancy at birth was 70 years or more!

Although some deficiency in relation to energy norms or to standards of growth in children need not indicate any serious long-term consequence, there appears to be a threshold beyond which such deficiencies can indeed be catastrophic. Lincoln Chen et al. (1980) show on the basis of their study of infant and child mortality in a district of Bangladesh that the death rates of children classified as "normal," "mildly malnourished," and "moderately malnourished" were not significantly different. But "severely malnourished" children had significantly higher mortality rates—about four to six times larger than for all other boys and girls of the same age. This study and others quoted by Eberstadt suggest that creditable performance in terms of life expectancy at birth or infant and child mortality are due to factors other than the nutritional status of the relevant populations as measured by FAO/WHO or other Western standards. It will be argued later on that these other factors are related to poverty and underdevelopment.

Before concluding this section, it is worth pointing out an important problem with the data on which many of the estimates of the extent of undernutrition are based. Even though undernutrition was defined as a pathological state of an individual arising from an inadequate intake of food over a *considerable period of time,* the available estimates of global extent of undernutrition are not based on *longitudinal data* on intakes of food of a suitably chosen sample of *individuals* from the relevant population groups. Nor are they obtained from anthropometric or clinical and biomedical evidence from such samples. The estimates are based neither on individual *intake* data nor on data on supposed outcomes of inadequate intakes such as anthropometric or clinical evidence. Very few sample surveys of individual intakes exist and even these few provide data on intakes during a *day* or at the most a *week.* Even rarer are surveys that provide data on body weight and activity pattern of surveyed individuals. Strong and untested assumptions will be needed to interpret these intakes as *habitual* or *long-term* intakes of the individuals surveyed. Most consumption surveys provide data on *expenditure on food* at the *household level* over some reference period such as a week, month or even a whole year. Even if one were to set aside the myriad problems associated with obtaining complete and reliable estimates of consumption through questionnaires in developing countries where the components of consumption consisting of homegrown food, food obtained in barter and as wages in kind are significant, a whole host of assumptions of varying degrees of plausibility is needed to go from expenditure on food to estimates of energy intake at the *household* level. The data situation is even worse since surveys of consumption expenditures are not available for all developing countries and only for a few countries such as India are data from more than one survey available. Given this situation, the brave estimators of the FAO and World Bank literally make bricks without hay. With little more to go by than an unreliable food balance sheet for a country as a whole and even less reliable data on per capita income, they arrive at estimates of the extent of undernutrition by making heroic and often untested and even untestable assumptions about the distribution of income and food among individuals in the population.

Tables 1 and 2 present data on the frequency distribution of households in rural and urban India in 1971-1972, according to calorie intake per consumer unit.[5] If the FAO/WHO norm (2,700 kcals per day per consumer unit) and the theory underlying it were correct, a large proportion of Indian households must be under severe nutritional stress—at the lower end of the spectrum due to extremely inadequate intake and at the other end due to excessive intakes. There is no clinical or biomedical evidence to support such an inference. This is not the occasion to enter into a detailed discussion of the survey and the data.[6] It suffices to say that given the Indian socioeconomic environment, there are strong reasons to believe that the consumption of the poorer households is understated and that of the richer households overstated in the survey. Yet the Indian surveys are some of the best of those from developing countries.

An equally striking illustration of the dubious value of calorie intake norms comes from the analysis of Bhalla (1980). Using the data from a Health and Nutrition Examination Survey (HANES) of over 20,000 people conducted by the U.S. Department of Health, Education and Welfare in April 1971 and June 1974 and FAO/WHO norms, he shows that 67 percent of American males and 80 percent of American females have a calorie intake below requirements! No one will seriously claim that such a high proportion of the U.S. population is undernourished.

Production, Consumption, Trade, and Transfer of Food

Professor Poleman has given several reasons available estimates of food production in less-developed countries (LDCs) may be underestimated. Several other problems with production data may be worth mentioning briefly, using the Indian data collection system for illustrative

Table 1

Calorie and Protein Intakes in 1971-72

All India: Rural Areas

Number of Sample Villages: 4200

Energy Intake per diem per consumer unit (k.cals) (1)	No. of Sample Households (2)	Average Household Size (3)	No. of Consumer Units per household (4)	Average Intake per diem per consumer unit	
				Energy (k.Cals) (5)	Protein (gms.) (6)
Up to--1500	651	6.61	5.01	1221	35
1501--1700	455	5.94	4.77	1604	45
1701--1900	576	5.93	4.75	1801	52
1901--2100	762	5.86	4.71	2002	57
2101--2300	854	5.95	4.79	2203	64
2301--2500	947	5.65	4.51	2399	70
2501--2700	882	5.73	4.55	2599	77
2701--3000	1234	5.43	4.31	2842	84
3001--3500	1774	5.21	4.13	3227	96
3501--4000	1174	4.94	3.92	3717	111
4001 & Above	2159	4.35	3.47	5248	170
All Groups	11468	5.39	4.29	2724	76

Source: Government of India, National Sample Survey Organization, National Sample Survey Report
No. 238, Volume I, Appendix III, Table (0.0R).

Table 2

Calorie and Protein Intakes in 1971-72

All India: Urban Areas

Number of Sample Blocks: 4375

Energy Intake per diem per consumer unit (k.cals) (1)	No. of Sample Households (2)	Average Household Size (3)	No. of Consumer Units per household (4)	Average Intake per diem per consumer Unit	
				Energy (k.cals) (5)	Protein (gms.) (6)
Up to--1500	1028	6.20	5.03	1238	35
1501-1700	808	6.20	5.01	1604	46
1701--1900	1201	6.18	5.01	1800	51
1901--2100	1569	5.93	4.80	2001	58
2101--2300	1744	5.73	4.64	2201	64
2301--2500	2054	4.92	4.00	2400	69
2501--2700	2075	4.37	3.55	2598	75
2701--3000	2571	4.29	3.48	2799	84
3001--3500	2830	3.92	3.17	3184	94
3501--4000	1525	3.60	2.90	3720	108
4001 & Above	2054	2.84	2.24	5217	15
All Groups	19459	4.72	3.81	2699	81

Source: Government of India, National Sample Survey Organization, National Sample Survey Report
No. 238, Volume 11, Table (0.0U).

purposes. An estimate of crop output in a season is obtained in India by multiplying the area sown with that crop in each season by the yield per unit area in that season. As part of the system of levying and collecting taxes on land and crops, the data on crop area were obtained in those areas where the system was established. Prior to independence, the yield per unit area was obtained mostly by "eye-estimation"—a procedure in which the village revenue official assessed the potential yield of a standing crop relative to "normal" yield by inspection. Since independence, eye-estimation has been progressively replaced by "crop-cutting" estimates based on average yield from harvesting a random sample of cuts (plots) of land growing the crop. Thus the output estimate is the product of supposedly *complete enumeration* of crop area and a *sample average yield*. Because the coverage of the area reporting system and the extension of crop-cutting to yield estimation increased over time, a series of "adjusted estimates" allowing for this increase is also published. As of 1976-1977, the reporting area covered 92.5 percent of the geographical area. In 1978-1979, the crop-cutting method was used for yield estimation from 95 percent of areas under cereals and 57 percent of the area under pulses. For other crops the proportion of area covered by crop-cutting varied from a low of 45 percent of area under sesamum to a high of 96 percent of the area under groundnut.

For a few years after the establishment of the National Sample Survey (NSS) in 1950, independent estimates of area and yield were put together by the NSS, both based on sample surveys. In addition, an indirect estimate of foodgrain output was obtained from the estimates of consumption of foodgrains thrown up by the consumer expenditure survey of the NSS. The three estimates differed substantially, with the official estimate the lowest, the consumption-based estimate the highest, and the sample survey–based production estimate falling in between.[7] There was a tendency over time for the estimates to come closer together. However, the NSS abandoned crop area and output estimation in the 1960s and no recent consumption survey-based estimate is available, leaving only the official estimate.[8] The discrepancy between consumption survey–based and production-based estimates of output is not unique to India but has been observed even in advanced countries. It is claimed that the sign of the discrepancy depends on the level of income, with consumption-based estimates exceeding production-based estimates in poor countries. The reverse is the case in richer countries.[9] Although the basis of estimation (and hence the possible sampling and nonsampling errors thereof) is well known for countries like India, for other countries it is not so well known (nor is it documented in detail in the publications of the FAO). The comparability of food production data between countries and over time within some countries is of doubtful validity. This is not to say, however, that such noncomparability is more serious for data on food production than, say, for national income data. In any case, there are no other data available.

In its World Development Report the World Bank (1982) has focused on agriculture and economic development. Table 3 is taken from this report. It would appear that developing countries as a group successfully maintained a growth rate of just under 3 percent per annum in their food output during the period 1960–1980, a rate considerably higher than that in their own earlier history and in the history of now developed countries when they were in a similar stage of development. It is also seen from this table that rapid growth in population led per capita output to grow at a very modest rate of 0.4 percent per annum. By far the most interesting and in many ways disturbing feature is the regional disparities in growth performance. During the 1960s, per capita growth of food output varied from a negligible 0.1 percent per annum in Africa, the Middle East, Latin America, and southern Asia to 1.8 percent in southern Europe. In the 1970s, per capita output of food *declined* by 1.1 percent per annum in Africa and stayed unchanged in southern Asia. The variation performance is shown even more dramatically in Table 4.

The regional imbalance in growth performance reflects the relative contributions of growth in cultivated area and yield per unit area to the aggregate growth during the 1970s. In India, even though there was no statistically significant difference in the annual rate of growth of output of food crops between the pre–Green Revolution period of 1949-1950 to 1964-1965 and the later decade of 1967-1968 to 1977-1978 (2.87 percent and 2.3 percent respectively in the two periods), growth of cultivated area at 1.40 percent per annum accounted for nearly half the growth in output in the first period. In the second period, area growth declined to 0.38 percent per annum, thus contributing less than a fifth to the output growth (Srinivasan 1979). Since almost all the arable land has been brought under the plough, any further expansion of cropped area must come from increases in cropping intensity (i.e., through cultivation of more than one crop during the year on the same land). Sizeable increases in cropping intensity can come only through investment in irrigation and drainage, but even with such investment, it is extremely unlikely that cropped area can grow as fast as it did prior to 1964-1965. According to the World Bank, even though there is ample cultivable and as yet uncultivated land in the humid and subhumid parts of Latin America and sub-Saharan Africa, this is not where the most needy people are. Land reserves in the Mediterranean area and most of Asia (except Indonesia) are extremely limited. China has also cultivated all of its arable land (World Bank 1982, p. 59). An implication of this is

Table 3

Growth Rates of Agricultural and Food Output by Major World Regions
Excluding China

Region and Country Group	Agricultural Output				Food Output			
	Total		Per-capita		Total		Per-capita	
	1960-70	1970-80	1960-70	1970-80	1960-70	1970-80	1960-70	1970-80
Developing Countries	2.8	2.7	0.3	0.3	2.9	2.8	0.4	0.4
Low-income	2.5	2.1	0.2	-0.4	2.6	2.2	0.2	-0.3
Middle-income	2.9	3.1	0.4	0.7	3.2	3.3	0.7	0.9
Africa	2.7	1.3	0.2	-1.4	2.6	1.6	0.1	-1.1
Middle East	2.5	2.7	0.0	0.0	2.6	2.9	0.1	0.2
Latin America	2.9	3.0	0.1	0.6	3.6	3.3	0.1	0.6
Southeast Asia	2.9	3.8	0.3	1.4	2.8	3.8	0.3	1.4
South Asia	2.5	2.2	0.1	0.0	2.6	2.2	0.1	0.0
Southern Europe	3.1	3.5	1.8	1.9	3.2	3.5	1.8	1.9
Industrial Market Economies	2.1	2.0	1.1	1.2	2.3	2.0	1.3	1.1
Nonmarket Industrial Economies	3.2	1.7	2.2	0.9	3.2	1.7	2.2	0.9
Total World	2.6	2.2	0.7	0.4	2.7	2.3	0.8	0.5

Note: Production data are weighted by world export unit prices. Decade growth rates are based on mid-points of five-year averages, except that 1970 is the average for 1969-71.

Source: World Bank (1982), p. 41.

Table 4

Food Production and Availability by Major Regions of the World

Region and Country Group	Index (1969-71 = 100) of Food Production per capita in 1978-80		Per Capita Food Availability (k.cals/day) in 1977	
	Average	Range	Average	Range
Low Income Countries	106	Low: 41 (Kampuchea) High: 121 (Sri Lanka)	2238	Low: 1754 (Ethiopia) High: 2695 (Afghanistan)
Middle Income Countries	108	Low: 53 (Hong Kong) High: 157 (Syria)	2561	Low: 1929 (Thailand) High: 3445 (Yugoslavia)
Oil Exporters	Not Available	Low: 69 (Saudi Arabia) High: 139 (Libya)	Not Available	Low: 2624 (Saudi Arabia) High: 2985 (Libya)

Source: World Bank (1982), pp. 110-111 and pp. 152-153.

Table 5

Growth Rates of Agricultural Production

(percent per year)

Period	Africa	Far East	Latin America	Near East	All the Developing Countries
Past (1961-65 to 1980)	1.8	2.9	3.0	3.0	2.8
Projected: Scenario A					
1980-1990	4.2	3.7	3.8	3.6	3.8
1990-2000	4.3	3.6	3.8	3.7	3.7
Projected: Scenario B					
1980-1990	3.3	3.2	3.3	3.1	3.2
1990-2000	3.4	3.0	3.3	3.0	3.1

Source: FAO (1981), Statistical Annex Table 4.

Note: China is not included

that either cropped area has to be increased through investment in irrigation at increasing marginal costs in Asia or cultivation has to be extended to fresh lands in Africa and Latin America, with the output produced there being internationally traded. Of course, in addition to expansion of area, yields could be raised through the cultivation of high-yielding varieties (HYVs) and through input intensification, particularly by the use of chemical fertilizers. This again would require investment in infrastructure (particularly in transport and communications) and agricultural research and extension services. There is a further issue as to whether the investment has to come from the public sector in the form of direct investment and if so in what proportion. Some of the costs will certainly be financed by farmers out of their own resources, provided sufficient incentives are there.

Some idea of the feasible growth (and the resource needs thereof) in 90 developing countries during the period 1980-2000 is available from FAO (1981). Tables 5 and 6 have been taken from this publication. Scenario A is based on optimistic assumptions regarding overall as well as agricultural growth in developing countries; scenario B is based on more modest medium growth assumptions. It is obvious from these

Table 6

Gross Investment Requirements, 90 Developing Countries

	A		B	
	1990	2000	1990	2000
		$ billion 1975		
1. Development of new land soil, water conservation and flood control	4.8	5.3	4.0	4.5
2. Irrigation (new and Improvement of existing)	12.7	14.6	10.5	12.1
3. Agricultural machinery (tractors and associated equipment)	14.9	36.0	12.0	26.1
4. Draught animal equipment and hand-tools	4.0	4.3	4.0	4.3
5. Work capital	4.2	8.3	3.2	5.3
6. Livestock (herd increases, and development of grazing land, meat and milk production investment including commercial pig and poultry)	14.3	21.6	10.1	13.8
7. Other	3.0	3.3	2.7	3.0
Total of primary sector	57.9	93.4	46.5	69.2
8. Storage and marketing	5.2	7.2	4.5	5.8
Total of OECD/DAC "Narrow definition" of agriculture	62.8	100.6	50.1	73.9
9. Transport and first stage processing	21.2	31.3	18.6	25.4
Grand Total	84.0	131.9	68.7	99.3

Source: FAO (1981).

tables that not only are massive investments called for but the geographical pattern of growth also has to be drastically changed compared to the period 1965–1980, even for achieving the more modest targets of scenario B.

A few remarks on fluctuations in food output around trend growth are in order. Gale Johnson (Chapter 1) has pointed out that the trend in world grain production since 1970 has been remarkably stable, with negative deviation from trend at less than 3 percent and positive deviations around 5 percent or less. In his view, political interventions in national and international foodgrain markets have been primarily responsible for price fluctuations and food insecurity. Since the issues of price stabilization and food security are to be discussed in other papers,[10] it suffices here to raise a few questions regarding fluctuations in output. It has been suggested that a step-up in the trend rate of output growth will bring larger fluctuations around the trend in its wake. An early proponent of this view was S.R. Sen (1967).[11] It is further claimed that the fertilizer-intensive new technology of production, while raising average yields, also increases its variance, and possibly the coefficient of variation as well. One of the arguments advanced in support of this view is that increases in trend rate of growth have to come in part from extension of cultivation to marginal and ecologically fragile lands and that this will inevitably increase fluctuations in yield and output. Further, it is alleged that the rapid spread of a few high-yielding varieties of seed over large areas carries not only the possibility that any pests and diseases to which these varieties are susceptible may inflict substantial and widespread damage, but also the even greater potential danger that the available genetic pool for further breeding will be progressively narrowed. There appears to be very little analysis and empirical work testing these views. However, if they are valid, a substantial increase in trend growth in world grain output in the future may be associated with an increase in fluctuations as well, even though such fluctuations in the past have not been of any serious consequence.

Hunger, Poverty, and Underdevelopment

I have argued that calorie requirement–based estimates of under-nutrition and poverty have no scientific basis. A pragmatic approach to identifying subgroups in the population with inadequate food intakes (not necessarily the same as the clinically undernourished group) is to consider several related aspects of food: share of food in consumption expenditure, marginal propensity to spend more on food due to increased total expenditures and composition of food consumption, in particular the shares of starchy staples, fats, meat, fish, and eggs.[12] At very low levels of per capita income (or total consumption expenditure), a household spends a very high proportion of its income on food and perhaps an even greater proportion of any increase in income. At such low levels of per capita income, the average propensity to spend on food will be close to unity, the marginal propensity will exceed the average, and most of the food consumed will consist of starchy staples. As income increases, the average propensity to spend on food reaches a maximum equal to the marginal propensity at that level of income and then declines. In some countries, even households with the lowest levels of income have sufficient income to be in the declining segment of the average propensity–income curve, so that the increasing segment of the curve is not observed.

Tables 7 and 8 present data from sample surveys of consumption expenditure in India. It would appear that in rural areas a *sustained* decline in the share of food in total expenditure and the share of starchy staples in calorie intake starts from a per capita monthly expenditure of 43 rupees (Rs). In urban areas, the corresponding figure is Rs 34. Interestingly, the rural cutoff point of Rs 43 per capita per month happens to equal the official poverty expenditure of Rs 15 per capita per month at 1960-1961 prices—a poverty line that has gained authority through its use in several studies of rural poverty in India (Ahluwalia 1978; Srinivasan and Bardhan 1974). This pragmatic approach leads to classifying around 42 percent of rural households and a little over 9 percent of urban households as having inadequate food intakes to *some degree*. To avoid any misunderstanding, it should be emphasized that one *should not infer* that these proportions represent households whose members are malnourished in a clinical or biomedical sense. Inadequacy of food intake in this approach is as perceived by the household and reflected in its consumption pattern. In any case, almost by definition, inadequacy of food intake is associated with poverty.

It is instructive to dig a bit further and analyze the incidence of rural poverty among different occupational groups. Table 9 is reproduced from Sundaram and Tendulkar (1982). It is seen that nearly 60 percent of the agricultural labor households are poor and account for over 44 percent of all poor households in rural India. It is not coincidental that these households suffer substantial bouts of unemployment. Nearly 40 percent of other labor households are also poor. Though this is not shown in Table 9, the poor households that are self-employed in agricultural occupations consist largely (though not exclusively) of small and marginal farmers, tenants, and sharecroppers. Thus the phenomenon of rural poverty in India is primarily one of insufficient access to sustained and productive employment or insufficient access to land.

Having access to land can be defined as either owning land or operating land owned by others under lease or tenancy arrangements. Tables 10 and 11 present the data on access. Even though land is of a heterogeneous quality and the data in these tables could thus be misleading, the fact that less than 10 percent of rural households own more than 50 percent of the land is unlikely to be changed significantly even if one were to allow for variation in land quality. It is interesting to note from Table 11 that the proportion of households not owning but operating land is a minuscule 0.20 percent. This is not surprising, since those who lease out land look for lessees and tenants who themselves

Table 7

All India: Rural

Distribution of Consumption Expenditure

Monthly per Capita Expenditure Class (Rs.) (1)	Percentage of households (2)	Percentage share of food in total expenditure (3)	Percentage of Calories derived from	
			Starchy Staples (4)	Eggs Meat and Fish (5)
00 – 13	0.39	82.50	89.40	1.22
13 – 15	0.35	82.91		
15 – 18	1.06	83.58	88.17	1.70
18 – 21	1.92	83.47		
21 – 24	2.96	82.98	87.54	1.95
24 – 28	5.28	83.51	85.61	2.76
28 – 34	11.10	83.21	84.81	3.16
34 – 43	18.53	82.46	82.58	4.25
43 – 55	20.19	80.41	80.30	5.39
55 – 75	19.48	76.43	76.38	6.71
75 – 100	10.26	71.29	73.77	9.71
100 – 150	6.07	63.00	66.04	11.35
150 – 200	1.46	52.46		
Above 200	0.95	45.15		
All classes	100.00	74.89	82.43	4.56
Sample Size	15467		11468	

Source: (1) Columns (1) – (3) are from Sarvekshana, Journal of the National Sample Survey Organization, Vol. 1, No. 1, July 1977, pages S-13, and S-38. The data are from the period October 1973–June 1974.

(2) Columns (4) and (5) are from Report No. 238, Vol. I of the National Sample Survey Organization, Pages 15 – 24. The data are for the period July 1971 – June 1972.

Table 8

All India: Urban

Distribution of Consumption Expenditure

Monthly per Capita Expenditure Class (Rs.) (1)	Percentage of households (2)	Percentage share of food in total expenditure (3)	Percentage of Calories Derived from	
			Starchy Staples (4)	Eggs Meat and Fish (5)
00 - 13	0.13	53.01	87.09	2.03
13 - 15	0.15	81.56		
15 - 18	0.28	79.42	85.44	2.54
18 - 21	0.36	81.45		
21 - 24	0.85	80.06	84.04	2.93
24 - 28	2.26	83.12	82.68	3.29
28 - 34	5.27	80.21	80.19	5.15
34 - 43	12.05	78.91	77.70	5.47
43 - 55	16.27	76.34	73.63	6.68
55 - 75	20.85	73.50	68.32	6.99
75 - 100	15.96	69.12	61.64	9.80
100 - 150	14.70	56.95	51.25	13.07
150 - 200	5.85	57.53		
Above 200	5.02	45.32		
All classes	100.00	67.72	71.08	7.43
Sample Size	7881		19439	

Source: (1) Columns (1)-(3) are from Sarvekshana, journal of the National Sample Survey Organization, Vol. 1, No. 1, July 1977, pp. S-77 and S-104. The data are from the period Oct. 1973-June 1974.

(2) Columns (4) and (5) are from Report No. 238, Vol. II, of the National Sample Survey Organization, pp. 1-6. The data are for the period July 1971-June 1972.

own some land, for land is the best collateral and security for ensuring that tenants do not default on rents or crop shares. Ownership of draught animals is also a prerequisite for obtaining land on lease, as there is no market for hiring draught animals in most parts of India. It is unlikely that a significant proportion of landless households own draught animals. The situation in Pakistan and Bangladesh is similar, as has been documented by I. J. Singh (1981). A World Bank study (1982, p. 78) states that "the great majority of the absolute poor—over 90 percent—

Table 9

Share of Different Types of Households in (a) All Rural Households
(b) Households Below the Poverty Line (c) Unemployed Persondays and
(d) Incidence of Poverty Within Each Household Type

All India (Rural): 1977-1978

Sr. No.	Household Type	Percentage Share in all Rural Households	Incidence of Poverty within each Household Type	Percentage share in all Rural Households below the Poverty Line	Percentage share of each Household type in Total Number of Unemployed Persondays		
					Males	Females	Persons
(1)	(2)	(3)	(4)	(5)	(6)	(7)	(8)
1.	Self-employed in agricultural occupations	46.11	30.37	34.97	20.74	11.44	17.61
2.	Self-employed in non-agricultural occupations	10.60	38.13	10.19	7.82	6.48	7.11
3.	Agricultural labor households	29.88	58.76	44.28	56.24	69.76	60.72
4.	Other labor households	6.88	38.54	6.69	11.10	9.82	0.68
5.	Other rural households	6.65	23.49	3.87	4.10	2.50	3.88
6.	All households	100.00	39.65	100.00	100.00	100.00	100.00

Source: National Sample Survey, Draft Report No. 293.

Notes: 1. Household types are defined by reference to the major source of
livelihood for the household in the year preceding the date
of survey.

2. All households with the per capita total consumer expenditure of
less than Rs. 50 per month constitute the set of poor households.
The poverty norm of Rs. 15 per capita per month at 1960-1961 prices,
when adjusted by the consumer price index of rural agricultural
laborers, works out to Rs. 48.45. We have approximated it at
Rs. 50 per capita per month.

3. Incidence of poverty within each household type in column (4)
is defined by the percentage of households below the poverty line
(of Rs. 50 per capita per month) within each household type.

are rural people who work on farms, or do non-farm work that depends
in part on agriculture. More than half are small farmers, who own or
lease their land; another 20 percent are members of farming collectives
mainly in China. The remaining one-fifth to one-quarter are landless,
and their livelihood is particularly precarious." Ensuring access to arable
land addresses only a part of the problem of poverty. In southern Asia,
where the substantial majority of the world's poor live, even in the
unlikely event of thoroughgoing land reforms, the resulting farm size
might not be adequate either to fully employ the family labor of the
peasant household or to provide adequate incomes without a change in
the technology of cultivation. It is true, however, that the seed-water-

TABLE 10

Access to Land in Rural India

Size Class (acres)	Percentage of households Owning Land in 1971-72	Percentage of Area Owned in 1971-72	Percentage of household Operating Land	
			1971-72	1974-75
0	9.64	0	27.41	3.08
.01 - 0.49	27.78	0.69	8.04	33.96
- 0.99	7.45	1.38	6.89	6.36
- 2.49	17.75	7.69	17.94	18.78
- 4.99	15.49	14.68	16.44	16.33
- 7.49	7.89	12.70	8.66	8.65
- 9.99	4.05	9.22	4.28	3.58
-14.99			4.72	4.28
15 & above	9.95	53.64	5.62	4.97
All Classes	100.00	100.00	100.00	100.00

Source: National Sample Survey Organization Report No. 215 and No. 288/1.

fertilizer technology of the Green Revolution is scale-neutral and, in principle, the small size of a farm does not bar its adoption. But for the full potential of this technology to be realized—particularly in the tropics—controlled irrigation and substantial inputs of fertilizer and pesticides are needed. In many parts of southern Asia there is either too much water during the monsoon season, so that it is difficult to control its use and ensure proper drainage, or there is too little water (particularly in the semiarid and arid tropics) for any irrigation. Even though plant breeders are developing high-yielding varieties for semiarid tropics, they are as yet not as widely diffused as the HYVs of wheat and rice. Farmers in such areas are placed at a disadvantage relative to farmers in those areas better endowed to reap the potential of the new technology. Regional disparities are bound to arise. If resources, including labor, are sufficiently mobile, the benefits *generated* in one area *accrue* to other areas as well.

TABLE 11

Ownership and Operation of Land in Rural India: 1971-72

Proportion of Households	Operating Land	Not Operating Land	Total
Owning Land	72.39	17.97	90.36
Not Owning Land	0.20	9.44	9.64
Total	72.59	27.41	100.00

Source: Same as in Table 10.

The use of purchased inputs such as fertilizers requires working capital. Often access to credit, particularly from commercial banks and government agencies, whether for working capital needs or for fixed investment in irrigation, farm implements, and draught power, is even more skewed than access to land. Traditional arrangements for the supply of credit, imputs, and marketing costs, however, such as the one in which a landlord provides credit to his tenant (sometimes even charging no interest) and the tenant agrees to sell his share of the crop to the landlord, need not be exploitative as they are often portrayed in the literature. This is not the occasion to expand on the rationale and efficiency of this and many other customary arrangements. Suffice it to say that in the absence of a well-developed, competitive and complete set of markets (spot and future) for inputs, outputs, and for risk bearing, traditional arrangements can be efficient. Haphazard state interventions in such a context will do more harm than good.

It has been alleged that the Green Revolution technology has led to excessive mechanization and displacement of labor and to alienation of land by small holders. Without getting into the vast literature on this topic, I will simply assert that the empirical support for this allegation is rather weak and the fault, if any, is not in the technology. Adoption of the new technology in a distortion-free context will benefit the poor peasants with small holdings and perhaps will even benefit the landless laborers by increasing the demand for labor. The careful study of I. J. Singh (1981) provides ample evidence that small farmers do not lag behind in the adoption of new technology and that the potential for improvement in their incomes as well as agricultural growth is substantial.

Landless agricultural laborers and self-employed rural artisans are often poor not only in the best of circumstances, but they are likely to be the most seriously affected whenever there is any unforeseen and serious decline in their purchasing power. In his study of famines, A. K. Sen (1981) shows that it was not the decline in available food that caused the Great Bengal Famine of 1943 or the more recent famines in Ethiopia or Bangladesh. In Bengal, military expenditure in urban areas and the consequent inflation in food prices were responsible. Wages of landless agricultural workers lagged considerably behind the galloping inflation, resulting in destitution of agricultural laborers, fishermen, and transport workers. Even though mortality figures by occupation groups are not available, there is no doubt that the mortality rate was highest among these groups. Although it is true that most governments (and particularly the government in India today) organize relief works to provide employment and a minimal level of real purchasing power whenever a threat of famine arises, that relief could be too late in coming, as illustrated by the recent episodes of famine in Africa.

The level of development of a country, particularly its agricultural development, affects the extent of poverty and hunger through its effect on employment opportunities of labor households and on the productivity of small farmers. It also means that safe drinking water, sewerage, and health facilities are inadequate, particularly in rural areas. Indeed, child mortality and growth retardation are due more to illnesses resulting from these factors than due to inadequate calorie intake. According to Padmanabha (1982), "Among infants, the major cause of death in the rural areas is tetanus as against prematurity in urban areas. Tetanus, pneumonia and dysentery, along with typhoid account for high proportions of deaths, among children up to three years. . . . These diseases are mainly conditioned by the absence or availability of basic facilities of reliable water supply, reasonable sanitation and basic services relating to child care" (pp. 1288–1289). Inadequate intake of nutritive food may weaken resistance to infectious diseases, but it apparently is not the major cause of mortality.

Another indicator of underdevelopment is illiteracy, particularly among women. In the same study, Padmanabha states that rates of mortality of infants born to illiterate mothers were 132 per 1,000 live births in rural and 81 in urban areas of India in 1978, as contrasted with 90 and 53, respectively, of infants born to literate mothers. These rates fall to 64 and 49 if the mother has a level of education beyond the primary school. These differences, though suggestive, are not conclusive because other factors—such as income—may have been different between the groups of mothers compared.

It is undoubtedly true that children and pregnant and lactating mothers run a far greater risk of nutritional stress in less-developed countries. However, it is somewhat arbitrary to assume, as Professor Poleman does in his paper, that the same proportion of these vulnerable groups is malnourished in all developing countries. Even though, as argued above, lack of access to sanitary, health, and educational facilities is a major reason for high rates of child mortality, available facilities are more equally distributed in some countries. Infant mortality rates in the state of Kerala in India illustrate this point. The infant mortality rate in Sri Lanka in 1980 was 44 per 1,000 live births, as compared to an average rate of 94 for all low-income developing countries (World Bank 1982, p. 150). The rate for Kerala was one-third of the rate of 136 for India as a whole. It is no accident that in Sri Lanka and Kerala literacy rates, particularly female literacy rates, are very high, and access to health facilities is well distributed. Some of the socialist countries such as China and Cuba have also achieved more remarkable improvements in health care than many other countries at comparable levels of per capita income. It is too sweeping to assume that the same proportion of infants and pregnant women are malnourished in all parts of the world. Indeed, it is futile and in many cases not particularly useful to come up with a global total of the hungry or malnourished. The set of policies needed to tackle the problem of hunger, whatever its extent and geographical distribution, will largely be specific to nations and locations. Global numbers are not particularly useful in formulating specific policies and may even be a hindrance in policy formation if by exaggerating the true extent of the problem they induce despair.

The above discussion focused on rural poverty because an overwhelming majority of the world's poor live in rural areas. But urban poor do exist, as is evident from the slums and shanty towns in most cities of the developing world. However, the very fact that they live in cities and urban areas means that they have better access than the rural poor to publicly provided amenities and services, including subsidized food, because such services and the public distribution of food through rationing are heavily urban-oriented. Perhaps the more relevant issues for urban poor are the availability of adequate shelter and steady employment opportunities. A development strategy that does not ignore the principle of comparative advantage and that does not penalize employment creation and reward overcapitalization will go a long way in tackling urban poverty.

The provision of subsidized (and sometimes free) food to urban residents is analogous to concessional sales and grants of food at the international level. The two may be connected in that the domestic distribution program may depend on concessional imports of food for its supplies, as was the case in India when P.L. 480 imports were the mainstay of distributions. Whether the supplies are obtained through imports or through purchases at below market clearing prices from domestic producers, there is the possibility of disincentives to domestic production. In designing price and distribution policies, the conflicting interests of producers and consumers have to be balanced. There is no need to delve deep into price policy issues here as they are discussed in Professor Reca's paper below (see Chapter 4).

Participating in international trade enables those countries that do not produce enough to meet domestic demand to augment their production through imports, and enables those countries that do to dispose of their surpluses. Such imports and exports are likely to take place even without fluctuations in domestic demands and supplies. However, fluctuations cause countries that produce enough on an average to meet domestic demands to export or import as needed, using international trade as an alternative to domestic buffer stocks for stabilizing domestic consumption. Of course, price stabilization policies (domestic and international) and the associated policies for stock accumulation and release—not to mention the financing of such stocks—raise a whole host of complex issues, as discussed in Chapters 4, 6, and 7.

Conclusions and Policy Implications

The much publicized estimates of the global extent of hunger and undernutrition by international agencies such as the FAO and World Bank have very little scientific basis. The methodology underlying such estimates, viz. an indirect procedure of arriving at the proportion of a population with energy intakes below requirements, is faulty for several reasons, the most serious being that human beings can and do vary their intakes within limits without any deleterious consequences to their health and activity. There is also evidence to suggest that an individual who is deemed malnourished either by the inadequacy of his energy intakes relative to "requirement" or by Western anthropometric standards is often not malnourished by clinical and biomedical assessment.

A more pragmatic approach to the definition of adequacy of food intake is to base it on more than one aspect of food consumption such as the share of food in total consumption budget, the marginal propensity to spend on food, the composition of food consumed in terms of starchy staples, meat, fish, eggs, dairy products, etc. Such a definition of adequacy of food intake is based on preferences, tastes, prices, and incomes rather than on some pseudoscientific notion of food requirement. As such, inadequate intakes in this sense *need not* lead an individual to become biomedically undernourished.

Not surprisingly, inadequate food intakes are mostly a reflection of poverty and underdevelopment. Poverty is predominantly a rural phenomenon and a large majority of the world's poor live in southern Asia, where the incidence of rural poverty is greatest among agricultural laborers, tenants, and small farmers. Lack of adequate access to land, irrigation water, and credit is a major cause of their poverty. Inadequate access to these productive resources precludes them from realizing the potential of the new seed-water-fertilizer technology. Apart from the lack of access to privately owned resources such as land, the rural poor (in fact, the rural population as a whole) do not enjoy to any significant extent publicly provided services such as safe drinking water, sewerage, facilities, health care, and education. Data from India indicate that diseases related to a lack of sanitary facilities and safe drinking water are the major causes of infant and child mortality, more than undernutrition caused by inadequate food. The mother's education seems to influence infant mortality. That better distribution of such services can reduce infant mortality and that it need not await substantial growth in incomes is shown by the remarkably moderate mortality rates in Sri Lanka and the Indian state of Kerala.

Several policies have been suggested, and a few have been tried, for alleviating hunger and poverty. In one view, there is no realistic hope of redistributing wealth or income, and only rapid economic development can help the poor through a trickle-down process. The opposite view is that without a radical redistribution of productive assets, particularly arable land, neither rapid and sustained economic development nor social justice are possible. Contradictory views have been expressed about the role of interregional (within a country) and international trade. Some advocate the allocation of agricultural resources according to emerging comparative advantage, even if it means specialization in the production and export of nonfood crops and import of food. Others advocate greater self-sufficiency in meeting the domestic demand for food, on the ground that foreign trade and aid in food have been used in the past to achieve political goals of donors and exporters. It is not in dispute that such use has been and is still being made, particularly by the United States. Whether this risk of political blackmail means that every developing country, however small, should strive to be self-sufficient at an enormous cost is arguable. It is becoming evident that in Maoist China the neglect of regional comparative advantage and emphasis on regional self-sufficiency in food imposed avoidable (and considerable) losses. Only recently is this folly being reversed.

Another argument advanced against following the dictates of comparative advantage is the one advanced by Lappe and Collins (1977) in their very readable book *Food First*. It holds that benefits from such a policy accrue to the rich, domestic and foreign, and that the costs are borne by the poor. They cite examples from Africa in which land the poor depended upon for sustenance was diverted by agribusiness to large-scale production of vegetables, fruits, and flowers for the European market. However, this reflects only the inadequacy of the domestic tax cum transfer system, and it does not mean that the principle of comparative advantage is faulty.

In addition to or in lieu of economywide policies such as those discussed above, several target group–oriented programs have been suggested or tried. These could be aimed either at increasing the food intake of target groups or at increasing their income-earning capacity. For example, food stamp programs, and subsidized public distribution of food, insofar as they are restricted to those considered to be poor, belong to the first category. So are programs, like the midday meal and the applied nutritions program in India, aimed at vulnerable groups such as children and pregnant women. School children are provided a free meal at school and pregnant women and infants are provided nutritional supplements, dry milk, etc. In practice, such programs have run into a number of problems. Leakage of benefits to nontarget groups occurred as, for instance, when the child being given a meal at school was no longer fed as much at home. Thus the school meal *substituted* rather than *supplemented* the home meal and in effect provided benefits to other nontarget members of the household. Ineligible individuals often get themselves included in target groups through corrupt means. Worse still, the targeting itself may be so bad, by design or accident, that many deserving poor get left out and some not so poor get in.

Examples of programs of the second kind are the Small Farmers Development Agency and a similar organization for marginal farmers and agricultural laborers in India. The intended purpose is to augment the earning potential of these groups through the supply of inputs at subsidized prices and to provide them with technical know-how. Although the potential for augmenting the production capacity of small farmers through investment in irrigation and the adoption of new technology and the use of chemical fertilizers is enormous, whether the currently operating programs will help achieve this potential is arguable. Access to the publicly provided benefits is often through several layers of corrupt bureaucracy located at towns away from the villages. The benefit to the poor peasant after the costs of several trips to the town, of lost time, of acquiring reams of documents and of course corruption may be negligible. Apart from this, such state interventions often end up by disrupting traditional production, credit, and exchange relations in the rural economy without replacing them with less costly and more beneficial arrangements from the point of view of the poor.

Attempts at ensuring food security through the operation of domestic buffer stock, coupled with the need for acquiring grain for the public distribution system in urban areas, have led to the introduction of several distortions in the foodgrain economy of many countries. To analyze just the Indian experience in this area would require another paper of equal length.

In concluding, a brief mention must be made of an interesting project under the leadership of Professor K. S. Parikh at the International Institute for Applied Systems Analysis at Laxenburg, Austria. The objectives of the project are to evaluate the world food situation and to suggest policy alternatives at *national, regional,* and *global* levels to alleviate food problems in the present and the forseeable future. Solutions to current problems that are consistent with a sustainable, equitable, and resilient long-term future are sought. The analysis is through a set of national general equilibrium models and a linking system to connect those models to form a global system. The particular strength of this approach is that it traces the full global general equilibrium effect of any policy variant. In one of the interesting counterfactual simulations, the impact of adding 30 million metric tons of wheat each year from 1977 to international trade was examined. As is to be expected the price path of wheat was lowered by the augmented supply. But interestingly, the short-term effect on prices was far more than the long-term one. Furthermore, a series of price-induced adjustments in production and trade took place with very little impact on hunger in the sense that not much of the extra 30 million m. tons of wheat a year reached the hungry. In almost all countries, an upward shift in the use of wheat as feed to increase production of bovine meat took place. An exercise with the India model simulated a program similar to one that used to exist in Sri Lanka, supplying to everyone 75 kg. of foodgrains per annum free of cost. It turned out that it would reduce GNP growth by a modest 0.8 percent per annum over the period of 1970-1990 while substantially increasing the food intake of the poor. These results are as yet preliminary, and a lot more work is to be done in testing the specifications of the models, yet they are enough to warn one that the general equilibrium implications of a policy may be considerably different (and may even be opposite) from its partial equilibrium effect. Parikh and Rabar (1981) provide a more detailed description of the project.

Acknowledgments

I wish to thank Robert Evenson for his comments and Lois Van de Velde for editorial assistance.

Notes

1. *Undernutrition* is different from *malnutrition*, which is defined as a "pathological state, general or specific, resulting from a relative or absolute deficiency or an excess in the diet of one or more essential nutrients. It may be clinically manifest or detectable only by biochemical or physiological tests" (FAO 1982). This distinction is not always recognized. Reutlinger and Selowsky (1976), FAO (1977) and even the present author (Srinivasan 1981) are guilty of ignoring the distinction and identifying undernutrition with malnutrition.

2. A reference man is between 20 and 39 years of age, weighs 65 kg., is healthy and employed for eight hours each day in an occupation involving moderate activity. He spends eight hours in bed, four to six hours in light activity and two hours in walking, active recreation, etc. A reference woman differs from a reference man only in her weight. She weighs 55 kg. Based on the "requirements" of a reference man and woman, those for men and women of other ages and activity levels were derived. Additional allowances for pregnant and lactating women were also specified.

3. This assumes that the distribution of requirements of the reference population is symmetric.

4. The study by Reutlinger and Alderman (1980) tries to rectify this by postulating a hypothetical joint distribution of intakes and requirements of individuals in a population. Since the "requirement" of a given individual is not a fixed number, this is yet another meaningless exercise.

5. A consumer unit is a male in the age group 20–39. Individuals of other age and sex groups have been converted to consumer units according to the following conversion factors:

	Number of consumer unit/s by age (in years)											
Sex	Under 1	1-3	4-6	7-9	10-12	13-15	16-19	20-39	40-49	50-59	60-69	70+
(1)	(2)	(3)	(4)	(5)	(6)	(7)	(8)	(9)	(10)	(11)	(12)	(13)
Male	0.43	0.54	0.72	0.87	1.03	0.97	1.02	1.00	9.95	0.90	0.80	0.70
Female	0.43	0.54	0.72	0.87	0.93	0.80	0.75	0.71	0.68	0.64	0.51	0.51

Source: Government of India, National Sample Survey Organization; National Sample Survey Report No. 238.

6. There has been a lively debate, initiated by Professor Sukhatme, in the *Economic and Political Weekly* of Bombay during the last two years on the use of FAO/WHO norms and Indian sample survey data in estimating the

extent of undernutrition. His views and those of his critics in this debate are available in Sukhatme (1982).

7. The government of India was sufficiently concerned about the discrepancies between official and sample survey estimates of output that it appointed a technical committee to study the problem. This committee even went into the question of whether the shape (rectangular or circular) and size of the sample cut affected bias and variance of the yield estimates. See India (1968).

8. Agriculture is under the jurisdiction of state governments. States with surplus food had to procure foodgrains for the public distribution system in deficit areas; there was a natural tendency on their part to understate their food output and for all states to exaggerate their food requirements. In some years, the officially published output estimate by the central government was politically negotiated with the states!

9. Professor Philip Payne of the Institute of Tropical Medicine in London suggested this in conversation. Apparently Japanese data covering prewar and postwar years show that as the Japanese economy grew richer, consumption based estimates went from above production-based estimates in the prewar era to below sometime in the postwar era.

10. For a comprehensive treatment of many of the issues see Valdés (1981).

11. For more recent analyses of Indian data see Mehra (1981) and Hazell (1982).

12. On the use of share of food in total consumption expenditure as an indicator of household welfare, see Deaton (1981). Professor Poleman has advocated the use of the composition of calorie consumption as a welfare indicator in his paper (see Chapter 2).

References

Acheson, K. J., I. T. Campbell, O. G. Edholm, D. S. Miller and M. J. Stock, "A longitudinal study of body weight and body fat changes in Antarctica," *Am. J. Cl. Nutr.* 33 (1980): 972–977.

Ahluwalia, M. S. "Rural Poverty and Agricultural Performance in India." *J. Dev. St.* 14, 3 (1978): 298–323.

Apfelbaum, M., J. Bostsarron, and D. Lacatis. "Effect of Caloric Restriction and Excessive Calorie Intake on Energy Expenditure." *Am. J. Cl. Nutr.* 24 (1971): 1405–1409.

Bennett, W., and J. Gurin. "Do Diets Really Work?" *Science '82* (March 1982): 42–50.

Bhalla, S. "Measurement of Poverty: Issues and Methods," Washington, D.C.: World Bank, 1980. Mimeographed.

Chen, L. C., A.K.M.A. Chowdury, and S. L. Huffman. "Anthropometric Assessment of Energy-Protein Malnutrition and Subsequent Risk of Mortality in Preschool-Aged Children." *Am. J. Cl. Nutr.* 33 (1980): 1836–1845.

Dandekar, V. M., and N. Rath. "Poverty in India." *Economic and Political Weekly* 6, 1 and 2 (1971), Bombay. Later issued as a pamphlet with the same title (Bombay: Sameeksha Trust, 1971).

Deaton, A. S. "Three Essays on a Sri Lanka Survey." Working Paper 11, Living Standards Measurement Study. Washington, D. C.: World Bank, 1981.

Eberstadt, N. "Hunger and Ideology." *Commentary* (July 1981), 40–49.

Edholm, O. G., J. M. Adam, M. J. R. Healy, H. S. Wolff, R. Goldsmith, and T. W. Best. "Food Intake and Energy Expenditure of Army Recruits." *Br. J. Nutr.* 24 (1970): 1091–1107.

Edmundson, W. "Individual Variations in Basal Metabolic Rate and Mechanical Work Efficiency in East Java." *Ecol. Food Nutr.* 8 (1979): 189–195.

Ferro-Luzzi, A., N. G. Norgan, and J.V.G.A. Durnin. "Food Intake, its Relationship to Body Weight and Age; and Its Apparent Nutritional Adequacy in New Guinean Children." *Am. J. Cl. Nutr.* 28 (1975): 1443–1453.

Food and Agriculture Organization. *Energy and Protein Requirements.* Report of a Joint FAO/WHO Ad Hoc Expert Committee. (Rome, FAO): 1973.

————. *Fourth World Food Survey.* Rome: FAO, 1977.

————. *Report of the First Joint FAO/WHO Expert Consultation on Energy Intake and Requirements.* Danish Funds-In-Trust, TF/INT 297 (Den) Rome: FAO, 1978a.

————. *Requirements for Protein and Energy: An Examination of Current Recommendations.* Report by a group of consultants. Rome FAO, 1978b. Mimeographed.

————. *Agriculture: Toward 2000.* Rome: FAO, 1981.

————. *Background Papers for the Meeting of Consultants on the Fifth World Food Survey, 19-21 May, 1982.* Rome: FAO, 1982. Mimeographed.

Grigg, D. "The Historiography of Hunger: Changing Views on the World Food Problem: 1945-80." *Trans. Inst. Br. Geogr.* N.S. 6 (1981): 279–292.

Hazell, P. "Instability in Indian Food Grain Production." Research Report No. 30. Washington, D.C.: IFPRI, 1982.

India, Planning Commission. *Report of the Technical Committee on Crop Estimates.* New Delhi: Manager of Publications, 1968.

Lappe, F. M., and J. Collins. *Food First.* Boston: Houghton and Mifflin, 1977.

Mehra, S. "Instability in Indian Agriculture in the Context of New Technology." Research Report No. 25. Washington, D.C.: IFPRI, 1981.

Ojha, P. D. "A Configuration of India Poverty." *Reserve Bank of India Bulletin* 24, 1 (January 1970), pp. 16–27.

Padmanabha, P. "Mortality in India: A Note on Trends and Implications," *Economic and Political Weekly* 17 (7 August 1982), Bombay, pp. 1285–1290.

Parikh, K. S., and F. Rabar, eds. *Food for All in a Sustainable World: The IIASA Food and Agriculture Program.* Status Report SR-81-2. Laxenburg, Austria: International Institute for Applied Systems Analysis, 1981.

Presidential Commission on World Hunger. *Overcoming World Hunger: The Challenge Ahead.* Washington, D.C.: GPO 1980.

Reutingler, S., and H. Alderman. "The Prevalence of Calorie Deficient Diets in Developing Countries," *World Dev.* 8 (1980): 399–411.

Reutlinger, S. and M. Selowsky. *Malnutrition and Poverty.* Staff Occasional Paper No. 23. Washington, D.C.: World Bank, 1976.

Sen, A. K. *Poverty and Famines.* London and Oxford: Oxford University Press, 1981.

Sen, S. R. "Growth and Instability in Indian Agriculture." *Agricultural Situation in India* 11, 10 (1967): 2–16.

Singh, I. J. *Small Farmers and Landless in South Asia.* Washington, D.C.: World Bank, 1981. Mimeographed.

Srinivasan, T. N. "Trends in Agriculture in India, 1949-50–1977-78." *India Economic and Political Weekly*, Special Number (August 1979) Bombay, 1283–1294.

————. "Malnutrition: Some Measurement and Policy Issues." *J. Dev. Econ.* 8 (1981): 3–19.

Srinivasan, T. N., and P. K. Bardhan. *Poverty and Income Distribution in India.* Calcutta: Statistical Publishing Society, 1974.

Sukhatme, P. V. "Measurement of Under Nutrition." Maharashtra Association for Cultivation of Science, Pune, India, 1982. Mimeographed.

Sukhatme, P. V., and S. Margen. "Auto-Regulatory Homeostatic Nature of Energy Balance." *Am. J. Cl. Nutr.* 35 (1982): 355–365.

Sundaram, K., and S. D. Tendulkar. "Towards an Explanation of Inter-Regional Variations in Poverty and Unemployment in Rural India." Working Paper No. 237. Delhi School of Economics, University of Delhi, 1982.

Taylor, L. *Food Subsidy Programs: A Survey.* Report prepared for the Food Foundation, New York, 1980. Mimeographed.

Valdés, A., ed. *Food Security for Developing Countries.* Boulder, Colo.: Westview Press, 1981.

Widdowson, E. M. "A Study on Individual Children's Diets." Special Report Series, Medical Research Council, 257. London: HMSO, 1947.

World Bank, *World Development Report.* New York: Oxford University Press, 1982.

Discussion _____

Vinod Thomas
World Bank
Washington, D.C.

The issue of defining hunger and its causes is both crucial to the development task, and controversial. Given the complexities involved in defining and measuring hunger and in understanding its sources and causes, widely differing views on the subject have been expressed over time. The papers by Professor Poleman and Professor Srinivasan examine, inter alia: aggregate food supplies, their adequacy and implications for hunger, and selected policy implications. Both papers elaborate on the measurement procedures utilized in much of the previous work on estimating hunger. A central proposition raised in both accounts is that these conventional measures substantially overstate global hunger.

This view on the magnitude of world hunger presented by the authors does not fall into the mainstream of much of the previous technical work; it is also at odds with popular writings on the issue. Differences in quantitative estimates are enormous—a range of tenfold at the extremes. Given the central role of food in the development process of the countries concerned, such divergence in views is not helpful. My objective, therefore, is to seek a greater consensus on the subject by highlighting the agreement in some areas, and by bringing out the grounds for differences of opinion in others.

First, the majority of people would concur that food is about the most basic of human needs, and therefore its deprivation, where it occurs, constitutes an area for urgent policy action. Second, there is also a growing recognition that food shortage, regardless of its exact magnitude, is more a problem of distribution than of production. This is not to minimize the production problem faced by several countries, particularly in sub-Saharan Africa. Nevertheless, it remains true that a better distribution could solve any food deficit actually existing in the world today. According to the *World Development Report, 1980*,[1] for example, a redistribution of just 2 percent of the world's grain production could eliminate malnutrition; available grain output is also estimated to be sufficient to provide everyone with 3,000 calories and 65 grams of protein daily. Thus, we could agree that because food is a basic need, and because the means to solve hunger are available, the food problem, whatever its exact magnitude, should receive urgent policy response.

A third area for agreement concerns the importance of nutrition in human development. The positive contribution of improved nutrition to the growth and performance of children, for example, is well documented. It would seem to follow, therefore, that much can be gained from promoting nutritional education and other means for achieving better food balances, and from supporting adequate and proper food intakes, particularly on the part of vulnerable groups.

Although there is, or at least ought to be, agreement on the policy imperative to address hunger, the extent of the problem today is disputed, particularly in the papers by Poleman and Srinivasan. In this context, I would like to briefly discuss three areas where greater consensus would be desirable: the magnitude of hunger; the importance of food in poverty measures; and the implication of measures of hunger and poverty for policy.

The differences in the estimates of hunger are wide, ranging from a few tens of millions to over a billion. A set of lower-range estimates is provided by Professor Poleman; some higher ones have been developed in the past at international organizations working in this area. Two steps are involved in the measurement procedure usually adopted by the latter: the application of established norms for food intake, and the use of estimates of food availability for consumption. Poleman and Srinivasan both contend that these norms are overstated and the estimated availability understated, thereby overestimating the gap between the two that is the measure of malnutrition. On the other hand, it should be noted that the lower figures can probably be only defended as lower bounds on hunger, and not as accurate measures of the total size of the problem.

Professor Srinivasan sets out arguments for the lack of adequate scientific basis for the widely used calorie requirements. These norms are also devoid of sufficient behavioral underpinnings and they usually overstate the basic human needs for food. It may also be added that over the range of these norms, the income elasticity for food is often observed to be rather low, and the price elasticity rather high, neither of which characterizes food as a critical need at the levels of incomes they usually correspond to. Professor Srinivasan's inclination is to discard the use of norms and look elsewhere for indicators of food inadequacies.

An alternative I would support, however, would be to lower the norms to more meaningful levels, to ranges over which they can serve a better discriminatory role in identifying people facing serious food deficits. Supplemented by behavioral indicators of food adequacy such as the marginal propensity to consume, the proportion of food in the consumption budget, the switch from quantity to quality in food intake, etc., the use of low norms could possibly serve to identify hunger. More importantly, such yardsticks can be useful in highlighting spatial and temporal interpersonal differences in "requirements." In a recent study,[2] I have noted that measured nutritional deficits on aggregate vary widely for Brazil as a whole, depending on which of the available alternative norms is adopted. On the other hand, these norms—which vary somewhat spatially—also serve to bring about consistent differences in the nutritional status across locations. In particular, sharp disparities between the northeast of Brazil and the more affluent southeast are indicated. Rural areas, on the other hand, fare as well as or better than urban areas in nutritional terms.

Sources for estimating food availability include consumption surveys and national food balance sheets. Both authors note a variety of reasons— such as the failure to adequately account for on-farm consumption— why these sources consistently understate true food availability. It should be noted, however, that a tendency to overstate agricultural output may also exist in situations where local officials find it necessary to show evidence of good performance. Furthermore, where output is systematically understated, estimates of year-to-year variations could still be reasonable—as borne out by price data in many instances—even if the measured level for any year is too low. Nevertheless, the dominant tendency seems to be to understate food availability, which, combined with high norms, can lead to an overestimation of food gaps.

It would appear, therefore, that the room for inaccuracies under existing methods of estimating food deficits can be substantial. At the same time, it should be recognized that more satisfactory alternatives are yet to be developed and applied in any significant scale. These considerations also raise questions about the need for global estimates of hunger, and about their impact on policy formulation. It is difficult to delineate the positive effect past projections of global food deficits might have had on actual performance. It may be noted that in spite of the high estimated food shortages, food aid is on the decline. On the other hand, effects of hunger estimates on regional and sectoral investment decisions could probably be pointed out. Policymakers in developing countries often need to measure disparities in food consumption across locations; even if global and aggregative measures may not be particularly helpful, more refined estimates to bring out such differences across regions and areas could still be useful in policy formulation.

Extensive use is made of measures of hunger in the estimation of poverty. Available global measures of poverty are comparable to those of hunger. According to the *World Development Report, 1980*, about 780 million people may be considered to be poor. The similarity between estimates of hunger and poverty arise in part from the central role of food in almost all poverty measures. One measure of "absolute" poverty establishes the income level at which nutritional norms are typically met, and then reads off the number of people who have incomes below this determined poverty level. An alternative approach prices a basket of food that can typically meet minimum nutritional norms and in addition, makes allowance for some "minimum" nonfood consumption in estimating a poverty expenditure threshold. These and other approaches base poverty measures primarily on food intakes; although they allow for varying degrees of substitution in the choice of food items in consumption, the comparison with nutritional norms is a feature of most such approaches.

The identification of food as the major component of a poverty consumption basket is not surprising. If a single indicator of poverty were to be picked, most people would choose food intake. In addition to being a basic need, food constitutes the bulk of consumption expenditure of the poor and is positively associated with income in the case of the poor. On the other hand, the use of high nutritional norms would focus attention on income ranges where food intake, particularly in terms of calories, is no longer very sensitive to income changes. As a result, small changes in the norms, particularly in caloric terms, would be consistent with fairly large changes in the corresponding income levels associated with them, and hence with large variations in measured poverty. At these high caloric requirements, it has also been observed that in addition to a low income elasticity of demand for calories, their price elasticity is rather high—neither of which is consistent with the notion of "minimum" needs. With high caloric norms, therefore, food is unlikely to serve as an adequate and discriminatory indicator of poverty.

In fact, at the mean, differences in food consumption are generally far less than those in nonfood consumption across countries or within countries. Although the average per capita income for middle-income countries is about six times that of low-income countries, average caloric intake in the former countries is estimated to exceed the latter by only about 10 percent. In studies of Brazil and Peru, I have found that disparities in real consumption among regions and between urban and rural areas arise mostly from nonfood categories and public services;

in many instances, rural areas achieve comparable or better food consumption than urban areas. In the case of Peru,[3] it was found that while nominal expenditure on food in the Urban Coast was 76 percent higher than in Rural Sierra, estimated real difference (after accounting for price variations) was only 34 percent; for nonfood consumption, although price differences were sharper, real expenditure difference between the two locations was still estimated to be about 56 percent. In a similar comparison in Brazil, it was found that while estimated real expenditure on food in the state of Sao Paulo was 40 percent higher than in the northeast, the average difference in nonfood consumption was about 130 percent.

Although food is probably not the main source of mean differences in real expenditures, it would still explain much of the low consumption of the very poor people. In the Brazilian study a difference of over 50 percent in food consumption (adjusted for price differences) was found between Sao Paulo State and the northeast at the fortieth percentile of the income distribution. As in the case of hunger measurement, reasonably low norms could serve to bring out a main aspect of poverty. Difficulties in properly measuring the absolute magnitude of poverty would still remain—as in the case of hunger—but adequate comparisons of locations would still be possible, provided spatial price differences are properly accounted for. Most governments attach less significance to any single measure of aggregative poverty than to differences in its incidence across locations and over time.

In lowering conventional norms to somewhat more realistic levels, care should be exercised in thereby not excluding some of the poor from their definition. To ensure a proper accounting of the poor, other indicators of welfare such as life expectancy and the availability of piped water and sanitation should be increasingly relied upon. Nutritional status, as well as these other indicators, however, only represents various symptoms of poverty. In measuring these symptoms, therefore, it should not be overlooked that the more fundamental causes of poverty—lack of productive employment opportunities and lack of education and skills—deserve more attention. The most effective solution to hunger is to raise the incomes and employment of the poor, although this may remain a longer-term option. In the meantime policymakers will continue to face the need to tackle the symptoms of poverty, through admittedly short-term measures. Recognizing this shortcoming, the process of policy formulation could and should be assisted through careful economic evaluations of alternative means to achieve given goals. By now a substantial body of information is available on the merits of various types of target group–oriented programs, such as food distribution schemes. More systematic analysis of the effectiveness of such options needs to be carried out within the context of the developmental goals of the countries in question.

Notes

1. World Bank, *World Development Report, 1980* (New York, Oxford University Press, 1980).

2. World Bank, *Differences in Income, Nutrition and Poverty Within Brazil.* Staff Working Paper No. 505 (Washington, D.C.: World Bank, Feb. 1982).

3. World Bank, *Spatial Differences in Poverty: The Case of Peru.* Reprint Series No. 153 (Washington, D.C.: World Bank, March 1979).

Part 3

4

Price Policies in Developing Countries

Lucio G. Reca
Universidad Nacional del Sur
Bahía Blanca, Argentina

Agricultural price policy deals with a varied and complex set of situations. The level of agricultural prices relative to other products or sectors in the economy and the price relations within agriculture, including relative prices between products and the product-factor price ratios, are among the most important. Sometimes the expression "price policy" is used rather loosely, making it difficult to interpret what it is supposed to focus on. Nebulous as the term may occasionally be, price policy has the virtue of attracting fervent supporters as well as staunch critics. On occasions the defense of or the attack on price policy has gone beyond reasoning. At times price policy is regarded as a panacea—just let prices perform their functions and all problems will be solved—and at times it is discredited on the basis of its alleged inability to promote production and its regressiveness concerning income distribution. Probably there is no other tool of economic policy, with the exception of land tenure, capable of dividing lay and expert opinions as strongly as price policy.

This paper deals with the objectives of price policy, the instruments most frequently used to implement them, and the effects of price policy.

Objectives of Price Policy

Increase Production

The relationship between product prices and agricultural production stands at the center of economic analysis. For a long time the nature of this relation was subject to question. Back in the 1930s there were even references to a "perverse" supply function, implying that the agricultural sector would tend to increase production in the presence of declining prices, contradicting in that way what common sense would anticipate. Later on it was shown that this apparently anomalous behavior could be explained by taking into account the phenomenon of short-run reaction. Each farmer tried to increase his output as a way to compensate for declining prices, without noticing that the consequences of this behavior by all producers would push prices downward even more.

Immediately after World War II a school of thought that acquired importance in the developing world, particularly in Latin America, explained the functioning of the economy mainly in terms of structural constraints. The agriculture side of the structural school identified land tenure and a nonentrepreneurial attitude of landowners as major constraints to increasing agricultural production. Little room was then left for prices as important elements in the process of mobilizing resources to generate additional output. This conception of the functioning of the agricultural sector led to the postulation of structural reform and technological change as the key factors to promote agricultural growth.

The controversy on the nature of the price-production relationship—which was a vivid one in academic and political circles—has been illuminated by numerous studies that have consistently shown a positive relation between prices received by farmers and production. These studies were carried out in several countries and under different conditions of farm size, tenure arrangements, and annual or perennial crops. They systematically show that increases in product prices are in general associated in the short-run with modest increases in production. It could be expected that a 10 percent increase in product prices would be followed by a 2–4 percent increase in output in the following production period. This relation holds when the agricultural sector is considered as a whole and when all available factors of production are being used. If, for example, there is a sustantial amount of idle land, the increase in production in response to a given change in product prices could be higher.

One important qualification to this discussion is that it assumes that all other important determinants of the level of output remain constant or experience only minor changes. Input prices, the level of technology, availability of credit, weather conditions, and the institutional setting are the most important factors. If any of these change simultaneously with the modification in product prices—a frequent situation—the effects of changing product prices are obscured. Unfortunately, in the real world

TABLE 1

Argentina: Grain Output and Prices, 1971–1982

Period	Area	Output	Yields	Prices
1971/1973	18.4	20.8	1.13	109
1974/1976	17.6	22.7	1.29	97
1977/1979	19.1	29.1	1.52	99
1980/1982	19.8	30.2	1.53	79

Note: area in million ha, output in million metric tons, yields in MT/ha and prices in constant pesos of 1970. Index of prices (1970=100) includes main grain crops weighed according to their importance.

a situation in which all relevant factors except product prices remain constant is exceptional.

Table 1 may clarify this point. A casual examination of the figures in the table would suggest that increases in output and price changes did not move in the same directions, in contradiction with what has been argued above. But the information in the table is "incomplete" in the sense that there are no references to changes in other variables. In this particular case most of the expansion in production was the result of the introduction of double cropping in a large region of the best agricultural land in Argentina, where short-cycle wheat and soybeans were substituted for corn. Technological change was the driving force responsible for this change in production. To complicate things a little more it could be argued (with some reason) that the institutional change which took place in Argentina in the mid-1970s created favorable expectations in the farming community. Producers anticipated less government intervention and thus the adoption of an already well known and proven technique was accelerated.

The argument of the meager response of output to increases in the level of agricultural prices is symmetric, i.e., production will not shrink in the face of declining prices. This is correct as long as the remaining variables influencing output do not change.

Another facet of product pricing policy deals with prices *within* agriculture as opposed to the situation analyzed above, where the agricultural sector was taken as a whole. When several products compete for the use of the same resources (land, labor and capital), changes in the price relationships among them will be reflected in changes in the composition of output, as long as there are technical possibilities of substitution between different activities. The case has some importance for grains (an increase in the price of corn relative to wheat will expand the area planted to corn at the expense of the area on wheat). It also has some importance when there is substitution between crops and livestock. In this particular case the consequences of changes in relative prices are not symmetrical. It will take some time to visualize the effects of an increase in beef-cattle prices in terms of larger quantities of beef brought to market, because the amount of time required for beef-cattle production exceeds that of grain production. The consequences of the inverse situation (a rise in grain price relative to livestock) will be evident in a shorter period through increases in the supply of both grain and livestock products. The latter will last while cattle numbers adjust to the new price conditions through a net decrease in inventories. The reason to mention these easily perceivable consequences of changes in product prices within agriculture, in situations where there is scope for substitution, is that with some frequency price policy centers on a certain product or products, ignoring the side effects of those decisions on other commodities.

Changes in relative prices within agriculture will usually induce larger supply responses than those mentioned for the agricultural sector as a whole. Resources are reallocated among alternative uses, and move to the most attractive alternative. In this case production expansion of one commodity does not necessarily imply an increase in overall output, because production of other commodities will decline in response to declining prices.

The results discussed so far have dealt with the immediate response of production to a change in the price of the product. If producers visualize the change as a permanent one, with prices remaining for some time at the new level, there will be additional increases in production. In summary, the long-run response of production to a given change in product prices is larger than the immediate (short-run) response. Those differences are usually significant.

Transfer of Resources from Agriculture

The share of agriculture in the domestic product of LDCs is usually large. This circumstance, coupled with the conception of agriculture as a sector relatively insensitive to changes in prices, made agriculture an ideal candidate for furnishing resources to other sectors of the economy to contribute to their growth. The ways in which the sector is taxed to effect the transfer of resources vary greatly. So does the amount of the taxes. In recent times a wider recognition of the role of economic incentives on agricultural growth has induced some reconsideration of the economic consequences of taxing agriculture.

Stabilization

Wide variations in prices, either within a year or between years, are usually regarded as undesirable by policymakers. Stabilizing agricultural prices or reducing fluctuations of some high-priority commodities is another important objective of agricultural price policy. It is sometimes argued that the reactions of local producers to large variations in world prices, considering the time lag characteristic of agriculture, could multiply the effect of external price variations. Exporting countries interested in avoiding abrupt changes in the internal prices of the commodities they export (assuming these products are important components of the domestic diet) favor price stabilization schemes. Importing countries trying to decrease their dependence on world market supplies, stimulate domestic production and set restrictions on agricultural imports. Seasonal variations of prices are also regarded as undesirable. They are associated with (large) variations in consumer income, partly at the expense of producer income. The rationale for intervention in this case is to minimize price variations within the year. Prices will tend to drop at harvest time and start increasing once the harvest season is over. In this particular case, price intervention, to be effective, needs to be backed by a competent marketing agency—a condition not always easy to fulfill. When stabilization is chosen as an objective of price policy, attention is primarily paid to the degree of price changes rather than to the level of prices.

Income Distribution

Here it is convenient to distinguish between the explicit objective of a policy and its unintended results. Virtually all policies affect income distribution in one way or another. A rise in agricultural prices, for example, will increase the real income of producers, including those who are very poor. At the same time real income of consumers will decrease. Those particularly hurt by the price raise are the urban poor and landless workers. Here there is a conflict between medium/long-run objectives and short-run constraints. Higher prices will induce some increase of production in the short run but some additional time will be required to reap the benefits of the larger production stimulated by the higher level of prices. In the meantime, the situation of the poorest consumers could be untenable.

An export tax on a commodity that is also consumed domestically also brings about some redistribution of income from producers to consumer (provided that the domestic market is not insulated from the world market). The redistribution from producers to consumers will be largely determined by the fraction of total output consumed internally.

Realizing Potential Trade Advantages

If a country can influence world prices by regulating the volume of its exports of a certain commodity, if the country is not a price-taker in world markets, in the short run it would pay to regulate the volume of exports. This would induce a better allocation of resources domestically, plus generate some fiscal revenue. However, higher prices for importing countries resulting from this policy might increase world supplies and change the demand conditions faced by the country with market power. The period of time considered is also in this case a crucial parameter.

Conflicts Among Objectives

The different objectives of price policy are not necessarily consistent. For example, an export tax on grain will generate government revenue and increase consumers' real income. The distribution between revenue and consumers' income will be roughly proportional to the distribution between local sales and exports. But such a policy will have a clearly adverse effect on production and may negatively affect the willingness of farmers to adopt new technologies and hence decrease output in the medium run. This would affect the export surplus and the balance of payments. These kinds of policies have been frequently used in many countries, small or large, governed by civilians or the military, and with varying degrees of private or public sector participation in the overall economic process.

Maximum prices below the equilibrium level on a commodity not internationally traded will benefit consumers in the short run as long as they can buy it at the regulated price. Yet this policy will have an adverse effect on resource allocation: as production of the taxed commodity declines, resources that could have been used in its production are released to less productive uses elsewhere in the economy. The choice of policies is largely influenced by short-run considerations, which often are in conflict with long-run objectives: the conflict between the transfer of resources from agriculture and the long-run growth of the sector illustrates the point.

Instruments of Price Policy

This section reviews some of the most important instruments used to achieve the objectives discussed above. The distinction between instruments and their effects will not be at times as neat as might be desirable.

Administered Prices

Prices that depart from the level they would have reached in the absence of government actions are called administered prices. They include product prices as well as factors of production. One of the most common forms of intervention in products markets is through support (fixed) prices. This denomination covers a wide spectrum of possibilities, as support prices could be fixed either above or below the level they would have normally reached in the absence of intervention.

Support prices above market equilibrium could perform two functions that are not mutually incompatible: they could stimulate production and redistribute income in favor of producers. Support prices are advocated on the basis of removing uncertainty (the farmer knows the price a crop will be selling for, and in that sense support prices work in favor of a better allocation of resources. For prices to perform these functions farmers must rely on the application of the price support schemes—a situation not always found in the real world. Continuity of a program of support prices, the existence of adequate channels to deliver the product, and prompt payment are necessary conditions for a price support program to operate effectively. In some instances (Brazil, for example) support prices are used as guidance to allocate subsidized credit.

There are also examples of support prices that even though regularly announced year after year are empty of any practical meaning: absence of or inadequate infrastructure to stock the grain or prices well below market equilibrium are the reasons for their uselessness. To justify the existence of support prices in these conditions it is argued that they are intended to set a floor if, for example, a bumper crop causes a sharp decline in prices. These programs are seldom operative.

There are numerous ways to calculate support prices, and considerable time and effort have been spent to find satisfactory procedures. However, the issue is far from settled. Costs of production offer one of the broad avenues commonly used to determine support prices, or market prices may serve as an alternative basis.

When the cost-of-production criterion is chosen, important aspects have to be solved before the calculation is made: type of technology chosen, farm size, expected yields, remuneration to family labor, and the cost of land are among the most pressing questions. In addition, profit has to be included on top of the costs (independently of the way they were computed) to arrive at the support price. It all results in a demanding exercise full of difficult and at times inevitable arbitrary decisions.

Support prices determined according to the market price criteria usually take into account some kind of historical trend to which a correction factor (above one if the objective is to protect or promote the crop) is applied. Conceptually this second procedure is easier to manipulate. It is often criticized in agricultural circles on the basis of ignoring the "real costs of production."

In many instances the procedures actually followed to determine support prices are even more complex than those described. Costs of production, market information, or both are elements brought into a decision-making process where many other factors also participate: the results of the last crop, the ability of the treasury to finance a potentially large purchase of grain, international market prospects, and the demands of the different groups in the political spectrum are some of the factors that (in addition to the purely technical ones) play a major role in the determination of support prices.

A final issue is the treatment of transportation costs. Exporting countries adhering to support price schemes usually define support prices net of transportation costs, according to the economic doctrine. When support prices are part of programs promoting self-sufficiency, the tendency is to set them on a flat basis, the same price all over the country. This last procedure may add some extra costs to the program, depending on the spatial distribution of producers and consumers.

Input Pricing

Increasing production as an objective of price policy has been up to now discussed considering the response of production to product prices. On the other side of the coin are input prices. Here again, policy can decisively influence the direction and intensity of use of the input mix through incentives or subsidies.

A well-known principle in economic theory states that the degree to which a factor is used in a production process is inversely related to the price of the factors, expressed in terms of final product price at the farm gate. As the price of the factor decreases in terms of the product price, the number of units of the factor that can be purchased with one unit of output increases, and the quantity of the factor used will also increase.

Price policies may deliberately check product prices with the alleged purpose of controlling food prices—the cost-of-living argument, may raise input prices (fertilizer for example) in order to protect a local industry producing fertilizer (the infant industry argument), or intervene in both markets simultaneously. The effect is to artificially depress the demand for the factor of production used below its most efficient economic use.

Input demand in developing countries has proved to be price responsive. Short- and long-run price elasticities of demand for fertilizer range between 0.3 to 0.7 and 0.7 to 3.0 respectively. Government intervention in fertilizer markets has been widespread, and in general it has helped to increase fertilizer consumption, which in the 1970s grew at the remarkable rate of 10 percent per year, increasing the share of developing countries in total nitrogen fertilizer consumption from 33 percent of world consumption in 1970 to 41 percent in 1980. Similar figures hold for other types of fertilizer.

In some cases fertilizer was promoted on the basis of high subsidies. These subsidies served different purposes: on occasion they were to compensate farmers for taxes on their produce or alternatively to encourage the adoption of fertilizer, a complementary input to improved seeds. In recent times, increasing fertilizer costs and the effect of subsidies on government budgets has stimulated considerable rethinking on the question of fertilizer subsidies.

Another interesting example of input pricing policy concerns water. Water used for irrigation plays several important roles in agricultural production. It helps decrease uncertainty regarding crop yields, as adequate moisture is made available at approximately the right time. It allows farmers to make more intense use of some factors of production like fertilizer, which without contolled water availability is often a risky investment. Last but not least, irrigation makes it possible to cultivate dry land that could otherwise contribute only very marginally to production.

Irrigated agriculture is capital intensive: dams, canals, and drainage systems are expensive. On-farm investment includes the costs of leveling land, the construction of feeder roads, housing, and clearing land. One largely unresolved problem in the field of agricultural price policy deals with the pricing of water services. A standard recommendation is that water charges should recover the capital costs incurred in the construction of the irrigation and drainage schemes, plus operating expenses. However, this practice is rarely followed, and water charges stand somewhere between token payments and a fraction of the value required to cover investment and operating costs. Recommendations from international lending agencies to reconsider the pricing of water according to the criteria mentioned or any other economically valid alternative—for example, the marginal contribution of irrigated water to production—rarely find a positive reception.

The late 1960s and early 1970s witnessed some frustrated attempts to substitute technology (for example the use of a new or better input) for prices as a tool to promote increases in production. It was argued that a sufficiently intense flow of technology complemented by appropriate extension services would be enough to induce farmers to adopt new technologies. In this model, an outgrowth of structural thinking, prices counted very little. The argument ignored the value of profitability under current technologies as a factor capable of inducing the adoption of new forms of production. For example, a well-conceived program of livestock development in Argentina, financed by an international agency, only took off after years of being available to producers, when large increases in beef prices occurred in the early 1970s. This issue is now more properly understood at the decisionmaking level, but it is one thing to accept the validity of an argument and quite another to act accordingly.

Export Taxes and Import Restrictions

Export taxes on agricultural products have a long and rich tradition as an instrument of agricultural policy in developing countries. They are easy to collect, which creates an additional incentive for their use. Schemes of export taxes differ depending on the main objective of the tax: transfer of resources from agriculture, or price stabilization. In the second case schemes tend to be more flexible. If the purpose of the tax is instead to transfer resources, a frequent device is to impose a percentage tax on the export price of the product. The level of taxation varies considerably but is usually significant. The popularity and persistence of export taxes suggest that they are highly regarded as a policy instrument. Export quotas (used as a price stabilization device and to secure domestic supplies) create a difference between domestic and external prices, but they do not generate government revenue.

Exchange rate policy could play a role comparable to that of an export tax. The effects of this instrument are more easily perceived in a situation with multiple exchange rates. If agricultural exports are converted into local currency using an exchange rate that is only a fraction of the conversion rate used for other transactions, this procedure constitutes a tax on all agricultural exports.

Another possibility is a general overvaluation of the local currency in terms of foreign currencies. This policy was fashionable in some Latin American countries in recent times. It was justified as being the proper policy to control inflation and to allocate resources more efficiently in the medium run. The experience of Uruguay is interesting in this respect: Beef-cattle production is the major agricultural activity and the number one source of foreign exchange earnings for the country. Uruguay exports about two-thirds of its annual beef production, so the importance of export prices as determinants of farm income is decisive. Uruguay

TABLE 2

Domestic Cattle Prices, Exchange Rate, and
Export Price of Cattle, Uruguay, 1978-1981

Year	Cattle prices	Exchange Rate	Export price of beef[a]
	--(values in constant pesos)--		
1978	100	100	100
1979	165	79	171
1980	105	55	175
1981	80	48	181

Source: Ministry of Agriculture, Uruguay.

[a]Beef export prices are average for Argentina and Australia
in US dollars of 1977 purchasing power.

endorsed the policy of overvaluing the exchange rate around the end
of 1978. The government acted consistently and steadily in terms of
the objectives the policy was supposed to achieve.

The effects upon cattle prices are shown in Table 2. In 1979 the
domestic index of cattle prices was close to the index of beef export
prices: the overvaluation of the Uruguayan currency was only beginning.
In the next two years, when the foreign exchange policy was fully
operative and the exchange rate declined to one-half of its value in the
base period, the divorce between domestic and world prices was substantial. These results, looked at strictly from the point of view of the
rural sector, show a sizeable transference of income from agriculture to
other sectors of the economy. It is true that for some time consumers
will benefit from beef prices lower than what they would have been
without overvaluation. But the medium-run costs of the policy will in
all likelihood be very high: low product prices act as a barrier to reinvest
in the sector and to adopt new techniques of production. They impose
in this fashion an artificial ceiling to production increases required to
expand the export surplus, required in turn to finance other investments
in the economy. On occasions export taxes are not set as such. Instead
of a formal export tax there is a maximum (single) price on a commodity
to be delivered to the only authorized buyer, usually a government
agency or a government-owned corporation. The agency is in charge of

the export trade. The difference between export prices and domestic prices is accounted for as agency income, but it is clearly an export tax.

There may be a whole panoply of import restrictions. These are used to protect agriculture and stimulate domestic production or to diminish the effects of external price variations on the domestic market. In some cases, import taxes are also used as a source of government revenue, but this is unusual. Import quotas play an important role in world agricultural trade. Their economic effects can be measured in terms of an equivalent tax, i.e., the percentage difference between the domestic price of a commodity when there is a quota and the price of the same commodity at the frontier of the country.

Marketing Boards

Marketing boards are the agencies in charge of executing important aspects of agricultural policy. They may buy, sell, or store grain when price stabilization is the objective of policy. In these circumstances they act as monopolists in the market or in cooperation with the private sector. Marketing agencies may take up export sales (and the purchase of food imports) under many different kinds of institutional arrangements.

Marketing boards frequently acquire a disproportionate importance in relation to other public agencies in the agricultural network. A powerful and wealthy marketing board coexisting with a weak and underfunded ministry of agriculture is a familiar feature in the developing world. Marketing boards are frequently financed with resources obtained from their regular operations, and are usually organized as autonomous agencies. These two characteristics give them considerable independence and initiative within the government. This is why marketing boards frequently evolve from policy executing to policy formulating agencies, in open competition with other units in the government.

Credit

Credit has often been used as a substitute for product pricing when the objective is to expand production. The rationale is that although it is undesirable to increase product prices, it *is* desirable to increase output. Cheap credit to cover operating costs and capital expenditures has been generously used. Inflation at times gives an additional benefit to borrowers on top of low and frequently negative interest rates.

Not all farmers have equal access to credit. Those who own land are in better position to secure credit. Those who have medium or large farms can also usually borrow under more favorable conditions than small farmers. These characteristics of credit markets have been analyzed extensively in the literature.

From the cheap credit policies typical until the mid-1970s, part of the developing world moved to quite the opposite extreme. As a consequence of macroeconomic policies, interest rates turned positive and high. Financial costs have become a major expense. In these circumstances, credit ceased to substitute for higher product prices in the efforts to expand production.

Effects of Price Policy

Price intervention in product or factor markets has the explicit objective of modifying a situation regarded as undesirable and potentially improvable. This section reviews some of the effects of governmental price policy on the agricultural sector.

Resource Allocation

Intervention in the pricing of products and factors is an incentive to shift resources to the most profitable alternatives, from the private point of view. Resulting changes, in turn, may have significant consequences for the contribution of agriculture to economic growth, as illustrated by the following example.

In 1979–1981, Mexico followed a policy of large increases in grain and other foodstuff prices at the producer level. Nonfood agriculture prices did not receive price incentives at that time, but inputs were generously subsidized. The combined effects of these interventions drastically changed the profitability ratio between the two groups of crops. Consequently, in some of the costly irrigated lands in the northern part of the country corn replaced cotton. This shift was a rational response from the point of view of private producers to the new price conditions: taking the difference between gross income and purchased inputs as a rough measure of profitability, cotton yielded 20 percent more than corn, a difference that probably did not compensate for the additional entrepreneurship required by cotton farming. If the same relationship between corn and cotton is computed excluding subsidies it comes down to 0.3.

The implication of this policy, in terms of resource allocation and contributions to total national output, is that policymakers gave an extremely high value to achieving self-sufficiency in food production. A measure of the cost of such an approach is given by the income foregone in the substitution process, production valued at nonintervention prices. Resource allocation is also distorted when price policy programs focus on a particular commodity without taking into consideration the existence of substitutes in consumption for the product in the program.

Guyana, for example, has consistently imposed a heavy tax on rice, in the order of one-half of the export price of rice. Intervention in this case is justified in terms of the importance of rice in domestic consumption, and the need to supply the population with cheap rice. Simultaneously the government is interested in the development of other food crops for which the country has some potential. The objectives (low rice prices at the consumer level and promotion of other crops) are largely incompatible. As long as there is some substitution in consumption between rice and other foods, low rice prices would artificially depress the demand for other crops and discourage the movement of resources to those activities.

Results like these do not require the taxation of domestic produce. As long as there is an effective subsidy on the consumption of an important crop, the development of other sources of food is slowed. Brazilian wheat policy provides a good example: Wheat production has been heavily subsidized for years through preferential product prices and cheaper investment and operational credit. On the other hand, wheat flour at the consumer level has been fixed well below the import price of wheat (which is in turn a fraction of wheat domestic price). The combination of this double subsidization policy lays a heavy burden on the treasury. On different grounds it puts a ceiling on the development of other food crops, of which corn is probably the most important: The price of wheat relative to corn declined by some 30 percent at the consumption level between 1974 and 1978. In the same period the consumption of wheat products grew at an annual rate of 5 percent. Considering two possible assumptions regarding the magnitude of the substitution of wheat for corn, the aforementioned decline in the price of wheat caused a reduction in the demand for corn in the order of 10–20 percent. The reduction in the price of corn resulting from this artificial reduction in demand (assuming a relatively inelastic supply function for both grains), would be in the order of 11 percent to 15 percent, which, in turn causes an annual reduction in the production of corn of 6 percent to 10 percent. These figures represented losses of potential output of corn in the order of 1.5 million metric tons per year, a figure close to Brazilian exports of corn in the early 1970s. The magnitude of the foregone output illustrates the effects of product price policies centered on a single product that don't take into consideration how the policy affects other components of the food supply.

Effects of Simultaneous Interventions

Government intervention in product and factor markets often takes place at the same time, leading to complex interplays between subsidies and taxes. The final effect of the returns to the direct factors of production in agriculture (mainly land and labor) will depend on the magnitude and sign of the price interventions. It is conceivable that a tax on the price of a product could be more than compensated by input subsidies. It is also possible that agriculture would be subject to a double squeeze from artificially low product prices and expensive inputs. These are empirical questions, and answers to them have come from the analysis of particular cases. The experience in Latin America largely suggests that price intervention in product markets is the decisive factor.

In Mexico (1981) the difference between gross income and the value of purchased inputs for corn produced in good rain-fed conditions, with grain and inputs subsidized, was 6.2 times what it would have reached in the absence of both types of subsidies. If input subsidies alone were removed, the difference only dropped to 5.8, while if only the subsidy on corn prices were removed the return to nonpurchased factors would have been 1.4 times the value of the nonintervention situation. Even though this last figure is indicative, in absolute terms, of a sizeable intervention, the bulk of intervention clearly lies on the price of the product.

Rice production in Guyana offers another example of a situation characterized by ample market intervention. One structural difference between rice in Guyana and corn in Mexico is that in Guyana rice is an exportable good, generating some 10 percent of the country's foreign exchange earnings. The other important difference is that rice is heavily taxed, as explained in the discussion of price policy instruments. An analysis similar to the one mentioned for Mexico shows, again, that the burden of price intervention is concentrated in product (paddy) pricing: in 1980 the difference between gross income and purchased inputs per acre of paddy was 0.38 of the level corresponding to a free market situation. A surprising result, in this case, is that when all factor market interventions are added up, direct subsidies on fertilizer and seed are cancelled out by the high cost of machinery services. Altogether, the cost of purchased inputs under intervention is some 8 percent higher relative to a hypothetical situation with no intervention in factor markets, an indication of the failure of a program aimed at compensating the effects of artificially lowered product prices through subsidies on inputs.

Demand for Inputs

The immediate effects of a price policy discriminating against the use of an input via artificially high prices restricts its use and generates some costs to society in terms of welfare losses. Furthermore, by reducing the demand for the input, the size of the market is artificially reduced, precluding the benefits from potential economies of scale. Policies that have artificially reduced the market for fertilizer illustrate this point.

Research policy could also be misdirected by the presence of distorted price ratios in the agricultural sector. If the cost of fertilizer is inflated, for example, fertilizer will play a secondary role in agricultural production. In that hypothetical circumstance, it would not be surprising if agricultural research paid little attention to developing crop varieties with the built-in capacity to benefit from a more intense use of fertilizers.

Irrigation water is frequently underpriced, with the following consequences: (1) Producers participating in irrigation projects under the conditions briefly outlined receive an additional benefit proportional to the difference between the marginal value of water to production and what they pay as water charges. Furthermore, if the land is privately owned, land values will eventually capture the extra income generated by the underpricing of water; (2) From a macroeconomic point of view, the inability of society to recapture the resources invested in a given project makes these resources unavailable for other investment projects, negatively influencing the possibilities of economic growth; and (3) Water underpricing may lead to overuse of water with undesirable consequences for resource conservation and extra costs associated with additional drainage requirements.

Production of Affected Crops

This is probably the most obvious effect of price policy. Following the positive relationship between product prices and output, increases in prices will lead to expansions in production and vice versa. The magnitude of the gains in output will depend on several factors like the nature of the product, the period of time allowed for the price to exert its influence, the number of products involved in the price programs, input prices, and the possibilities of technological innovation.

Annual crops will adjust to changes in prices more rapidly than perennial crops or livestock, a consequence of the biological characteristics of each. The period considered in evaluating the results of a price program in terms of production increases is a crucial question dealing with the relation between the short- and the long-run movement of resources between alternative products and between the agricultural sector and other sectors of the economy. The time necessary to complete the adjustment to a certain change in prices will vary according to the particular situation considered (commodity chosen, institutional, organization, biological characteristics) but will, in general, be several years.

As indicated above, the production effects of a price change on agricultural output as a whole will be much more moderate than on a particular commodity. The latter can benefit from faster reallocation of resources and does not face the limitation imposed by a fixed supply of land (assuming all cultivable land is being reasonably used).

Time and again higher product prices as an instrument to promote production have been discredited by meager production responses. There is some truth in this argument, sponsored, among others, by structuralist economists. Without a flow of new technologies, and if the aggregate elasticity of supply is low, higher agricultural prices will induce an income transfer from other sectors of the economy to agriculture, which may very well be more important than the production increases resulting from higher prices. This point illustrates the transcendence relationship between technological and price policy. This discussion has assumed continuity of the price programs and credibility by the farming community—two conditions not always present.

Income Distribution

The income distribution effects of price policy are ever-present and usually important. A few of them will be analyzed at this point:

Export Taxes. The effects of export taxes on income distribution are complex. In all cases they cause a transfer from producers to other sectors of the economy. Producer surplus is reduced in proportion to the amount of the tax. Domestic consumers benefit from lower after-tax prices. They experience an increase in welfare roughly proportional to the decline in prices and the amount of their consumption. Part of the surplus lost by producers is seized by government as export tax proceedings. Finally there is an economic loss in the process; a fraction of the surplus transferred by producers cannot be captured by any other economic agent. Usually this is a relatively small fraction. For example, the producer's surplus transferred as a result of a 40 percent export tax in a situation where one-third of total production is locally consumed and the remaining two-thirds are exported, assuming low elasticities of supply and of local demand and an infinitely elastic demand for exports, is distributed approximately in the following way: 30 percent increase in consumer surplus, 60 percent government revenue, and the remaining 10 percent is economic loss, a cost associated with the imposition of the tax. When the quantities of a product exported by a country influence world prices the effects of export taxation will include changes in foreign exchange earnings and losses in the welfare of foreign consumers in addition to the factors already mentioned.

Import Taxes. Import taxes have opposite effects to those discussed so far; consumers are the losers and producers and the goverment are the beneficiaries of the scheme. There is also an economic loss. In the case of agricultural products that are not internationally traded, the imposition of a tax will benefit the public sector (through tax collections) at the expense of consumers and producers. There will also be an economic loss for the system as a whole.

Input Subsidies. In addition to the macroeconomic effects, subsidies on inputs tend to have undesirable effects on income distribution between farmers, since most of the benefits are frequently captured by larger farmers.

Effects on Land Values. The effects of price policies will be also reflected in land values. When agricultural income increases as a consequence of government intervention in product or factor markets, land values will rise after a period of adjustment, which on occasions is very short. This is because land is a factor of production specific to agriculture and as such is a residual income claimant. Knowledge of the land distribution pattern will permit evaluation of a given program on the basis of its distributional impact.

Long-Term Growth

In addition to the specific evaluation of price policies, it is convenient to look at price policies from the point of view of their impact on the long-term growth of agriculture. Not all price policies have a positive impact on the long-run development of the agricultural sector. The effects of an export tax were discussed above. To that discussion it should be added that a persistent transfer of part of the producer surplus to other sectors of the economy will impair the ability of the sector to grow. In the same vein, artificially depressed product prices (as in the example of Uruguay) will be finally reflected in lower land values. In turn, cheap land acts as a disincentive to substitute other inputs for land and in that way it affects the long-term performance of agriculture. The problems of consistency mentioned in the section on objectives shows up here when the need for fiscal resources (obtained through an export tax) and the effects of tax on the long-run growth of the sector are considered.

Conclusions

This paper has attempted to review a broad and complex subject based on the experience of a group of less-developed countries. Individual experiences are not always comparable: differences in size, resource endowment, and institutional organization are too large to be ignored. With these limitations in mind, the main lessons that emerge from the use of price policy in Latin America in the last 20 years are:

1. There has been a broader recognition of the relation between economic incentives and agricultural growth. Even though price policies have not consistently recognized this important relation, agricultural policy in the last 10 to 15 years has been, in general, less biased against agriculture than it was in the 1950s and early 1960s.

2. Price policy probably attracts more attention than any other agricultural policy instrument. The multiple and far-reaching effects of price policy are still frequently exaggerated or ignored. Both attitudes create considerable difficulties in the design, implementation, and evaluation of the results in price policy actions.

3. There is widespread intervention in agricultural markets through price policies of different signs, intensities and duration. The examples of Mexico and Uruguay illustrate two different types of intervention.

4. Product and input pricing are the most visible components of price policy. They do not operate in isolation but interact with other instruments like credit, taxes (which are part of price policy if defined comprehensively), research, and land tenure. These interrelationships have not always been clearly understood. The result has been to create more confusion in connection with the use of price policy.

5. Export taxes remain one of the most important tools of agricultural price policy. Export taxes have been implemented directly (as a percentage on the export price of the product) or indirectly (through overvaluation of the local currency or through procurement at prices well below market prices). Uruguay and Guyana offer examples of the negative consequences of these policies.

6. Price support programs are widely announced in the region. They include food exporting as well as food importing countries. Criteria used to determine support prices vary and in large measure reflect the outcome of political processes rather than the results of a purely technical exercise. The effectiveness of these programs varies greatly. In some cases—like that of Mexico—they are backed by a well-endowed marketing agency. In other cases, support prices are meaningless (either the prices are too low or the marketing agency is not equipped to make the price good). Support prices for soybeans (Argentina, late 1960s) were partially responsible for the adoption of this crop by creating a market for it.

7. Continuity and reliability of price policy in its different versions continues to be a distant goal. Abrupt changes in policies (support prices, effective exchange rates, cheap credit) cause confusion and have a negative influence on resource allocation and long-term growth.

8. The conflict between achieving long-run or short-run objectives through policy has been resolved in favor of the latter more frequently than not. Price policy consistent with long-term growth continues to be relegated to second place. Fiscal needs calling for the imposing of export taxes and maximum prices on wage goods usually have top priority.

9. Price intervention in product markets has had more significant results than intervention in factor markets. These results hold whether inputs have been subsidized to compensate for controlled product prices

TABLE 3

Grain Production, Consumption, and Net Trade,
European Economic Community

Period	Production	Consumption (million m. tons)	Net Trade	Yields (m. tons/ha)
1960/1962	72.8	93.7	-20.9[a]	2.7
1970/1972	93.6	110.8	-17.2	3.5
1980/1982	116.4	116.7	- .3	4.3

Source: USDA, three-year averages.

[a] Negative numbers indicate net imports.

or whether product and input prices as well were subsidized. Guyana illustrates the first situation and Mexico the second.

10. Marketing boards in charge of price support programs have at times outgrown other units in the agricultural network, changing from policy executing to policymaking units and thus creating a confusion of roles in the agricultural public sector.

11. Price policy originates partially within the agricultural sector (for example, the level of support prices, decisions on subsidies for certain inputs, etc.). Ministries of agriculture and agricultural planning units have a leading voice in these areas. But decisions on other variables determining the effects of price policy on agriculture—like exchange rates and interest rates—are taken by other sectors of government usually alien to agriculture. This state of affairs is not the most desirable one from the point of view of an optimum use of price policy.

12. Price policies in the LDCs are not independent from policies in the industrialized world. The negative effects of concessional sales on agricultural growth in the developing countries has been clearly pointed out again and again, and in recent times new factors have emerged. First, the European Economic Community (EEC) has become self-sufficient in grains as a result of protectionist policies that have encouraged the adoption of land- and labor-saving techniques (Table 3). If the recent historical trends continue, the EEC may become a net exporter of grain, thus influencing production in LDCs, in the same way it became a net exporter of beef since 1980 at highly subsidized prices. Second, North America supplies most of the grain currently traded in the world. If as

a reaction to present economic conditions (low product prices and high financial costs) farmers increase production in a short-run effort to compensate for low incomes, the supply of exportable grain could be considerably expanded, negatively influencing production in grain-importing LDCs and competing with grain from grain-exporting LDCs.

References

Bale, M. D., and E. Lutz. "Price Distortions in Agriculture and Their Effects: An International Comparison." *Am. J. Agri. Econ.* 63 (1981): 8–22.

Behrman, J. R. *Supply Responses in Underdeveloped Agriculture.* Amsterdam: North Holland Publishing Co., 1968.

Brown, G. T. "Agricultural Pricing Policies in Developing Countries." in *Distortions of Agricultural Incentives*, ed. T. W. Schultz. Bloomington: Indiana University Press, 1978.

Brown, G. T. and G. Donaldson. *Agricultural Prices, Taxes and Subsidies: A Review of Experience.* World Bank Staff Working Paper (forthcoming).

Cuddihy, W. *Agricultural Price Management in Egypt.* Staff Working Paper No. 388, Washington, D.C.: World Bank, April 1980.

Hayami, Y., and V. W. Ruttan. *Agricultural Development: An International Perspective.* Baltimore: Johns Hopkins University Press, 1971.

Interamerican Development Bank. *Proceeding of the Seminar on Agricultural Policy: A Factor in the Development Process.* 17–21 March 1975. Washington, D.C.: Interamerican Development Bank, 1975.

Josling, T. "A Formal Approach to Agricultural Policy." *J. Agri. Econ.* 20, 2 (May: 1975): 175–195.

Krishna, R. "Agricultural Price Policy and Economic Development." in *Agricultural Development and Economic Growth*, ed. H. S. Southworth and B. F. Johnston. Ithaca, N.Y.: Cornell University Press, 1967.

Nerlove, M. "The Dynamics of Supply: Retrospect and Prospect." *Am. J. Agri. Econ.* 61 (1979): 874–878.

Peterson, W. L. "International Farm Prices and the Social Cost of Cheap Food Policies." *Am. J. Agri. Econ.* 61 (1979): 12–21.

Reca, L. G. *Argentina: Country Case Study of Agricultural Prices and Subsidies.* Staff Working Paper No. 386. Washington, D.C.: World Bank, 1980.

Schultz, T. W. "Constraints on Agriculture Production." in *Distortions of Agricultural Incentives*, ed. T. W. Schultz. Bloomington: Indiana University Press, 1978.

Tolley, G. S., V. Thomas, and C. M. Wong. *Agricultural Price Policies and the Developing Countries.* Baltimore: Johns Hopkins University Press, 1982.

Discussion _____

Malcolm D. Bale
Country Policy Department
The World Bank

Introduction

Although my primary purpose is to discuss Professor Reca's paper, I find myself in substantial agreement with the points he makes. I will, therefore, emphasize further some of the conclusions that he arrives at, embellishing the statements with examples from work done by myself and other colleagues at the World Bank. In order to stimulate discussion, my comments are presented in what I hope is a provocative and perhaps exaggerated manner.

Among agricultural economists it has become almost a cliché to note that government interventions in food markets on a large scale are a complex, costly, and pervasive practice in developing countries. Many books and scholarly articles have recently appeared on the subject (Schultz 1978; Tolley et al. 1982; Bale and Lutz 1981; Lutz and Scandizzo 1980). Yet on any day one can pick up the newspaper and read of further instances of intervention in agricultural markets. The *Washington Post*, for example, reported on 16 September 1982 that Argentina had just imposed a rationing system on several food items, including beef, and had fixed prices on fuel, bread, and milk. On the same day, the Philippines embargoed copra exports in order to encourage domestic coconut oil production. Agricultural economists have written extensively on the perils and pitfalls of agricultural price intervention policies, but the practice continues unabated. Why? Have economists failed to reach their ultimate audience—the policymakers—or are government officials

The comments made here are the author's and should not be attributed to the World Bank or its affiliates.

responding to a set of issues broader than economic dimensions that negate our economic reasoning?

The reasons given by policymakers for intervening in food markets are several: to achieve self-sufficiency in food supply; to maintain low prices on food items important in the cost-of-living index; to achieve internal price stability; to obtain government revenue; to ensure a constant income for farmers; and to keep food cheap for poorer urban consumers. Yet policies involving intervention in agricultural markets warrant further scrutiny by those who advise the governments of developing countries—not only because of the questionable validity of the intended objectives, but also because they frequently involve large financial and economic costs to the countries adopting the policies—costs often far in excess of the cost of government programs aimed at meeting other basic human needs.

The accepted attitude in many developing countries of governments playing a very paternalistic role in everyday economic life creates a favorable climate for many kinds of intervention in food markets. Usually, little thought is given to whether the intended benefits will actually be realized and whether alternative interventions could possibly produce better results at a lower cost. The pressures of the moment require that policymakers make quick decisions. The problem is that they respond to short-run imperatives with a continual series of short-run policy palliatives. The longer-run issues are seldom addressed and may in fact be exacerbated by the short-run measures. "Public policy decisions" becomes largely a euphemism for an incoherent sequence of desperate expedients. The net result inevitably is that scarce resources are misdirected, slowing the countries' rate of development and in turn impeding the achievement of a permanent solution to food production and distribution.

Types of Intervention

The number of policy instruments used in agriculture in developing countries is vast, being limited only by human ingenuity. Among those commonly found, the most prevalent are product price controls, export taxes, subsidies on inputs such as credit, fertilizer, and irrigation water, and monopoly parastatal buyers of farm output. In addition, the indirect tax or subsidy on agriculture of misvalued exchange rates is common.

In Africa, exchange rates are frequently overvalued, resulting in an implicit tax on agriculture of 10 to 30 percent. Tropical export products are further taxed, while basic cereal prices are held down by a variety of import and domestic consumer subsidies or by fiat (World Bank 1981).

In Latin America, substantial changes in agricultural policy that occurred during the mid-1970s have alleviated the problem of overvalued exchange rates. In some countries (Mexico) agriculture is receiving overcompensation for overvalued exchange rates while in others (Costa Rica) exchange rates are actually undervalued, providing agriculture with an implicit subsidy.

In Asia, policies differ substantially from one country to another. India limits exports of some crops (rice) but provides supports through a broad array of subsidies. The Philippines has similar offsetting production supports. Thailand has a variable tax on rice exports and intermittently imposes storage requirements on rice.

An interesting question arises from the observed behavior of countries simultaneously taxing agricultural output and subsidizing agricultural inputs. If the two policies largely offset one another, why do governments choose two policy instruments to achieve one goal when one instrument, such as a food subsidy, would achieve the same goal? There are three possible reasons. First, the two distortions (taxes and subsidies) may not and usually do not cancel out. Implicit taxes typically dominate. Second, it is difficult to design a food subsidy scheme that is fraud proof. Unless the commodity changes form, such as wheat to bread, the government becomes a milking machine, with the commodity being recycled through the government several times in order to get a multiple subsidy. Finally, the government may have multiple goals requiring several policies. The goals of food self-sufficiency and cheap food, accomplished by import restrictions and food subsidies, is an example.

Effects of Intervention

Economists are concerned about government interventions in markets because of the misallocation of resources that results from such actions. There are many effects generated by a single policy action. In the usual case in developing countries where agricultural output is implicitly or explicitly taxed, the static effects compared to an absence of price interventions can be listed as follows:

- Agricultural output is lowered;
- Food consumption is increased;
- Exports of agricultural products are decreased or imports of agricultural products are increased;
- Rural employment declines and rural to urban migration increases;
- There is a redistribution of income from the farming sector to the urban sector;

- Foreign exchange earnings from agricultural exports decline or foreign exchange requirements for agricultural imports increase;
- Land is undervalued and therefore not utilized in a socially optimal way;
- Usually government revenues collected from agriculture are increased; and
- A net efficiency loss (sometimes called a deadweight loss) is incurred.

Table 1 illustrates the dimensions of these effects for two products in three developing countries. As can be seen, the numbers are quite impressive. In Argentina, farmers receive one-half of the world price for their wheat and maize, and one-third less than the world price for their beef. In Egypt, farmers receive one-half of the world price for wheat and maize, and one-third of the world price for rice and cotton. Pakistan is not quite so extreme. Wheat is priced 20 percent below the world price and rice is priced 40 percent below the world price.[1]

Because of these pricing policies, wheat output in Argentina was estimated to be 4.7 million metric tons below what it would have been at world prices in 1976, while wheat exports were over 5 million tons below what they might have been, and Argentina lost $727 million in export receipts. In Egypt, production of wheat decreased by 520,000 tons per year and rice by over 2 million tons per year, while in Pakistan wheat and rice production were reduced by almost 1 million tons each.

The consumption picture is reversed. Developing countries consume more than they would in the absence of price intervention. Thus the policies clearly have a beneficial effect in providing more food for the nonagricultural population of the country. However, it is important to understand that this is achieved at the expense of the agricultural sector. It is also important to realize that the "truly needy" poor of developing countries are the rural poor or the formerly rural poor who have drifted into urban centers because of lack of rural employment opportunities. Rural people are poor, and rural people migrate to urban areas in part because farm prices (the prices of their product) are artificially low. A strong case can be made that if prices were deregulated each country's welfare would increase, and certainly agricultural output would increase. By insisting on subsidizing food for the urban population, many developing countries are opting for continuing impoverishment. Although policymakers in developing nations may not quite be in danger of killing the goose that lays the golden eggs, their policies certainly are lowering her fecundity.

Table 1

The Effects of Price Distortions on Two Staple Commodities in Three Developing Countries, 1976

	Decrease in production	Increase in consumption	Decrease in exports	Decrease in agricultural employment	Income loss of producers	Income gain of consumers	Increased government revenue	Decreased foreign exchange earnings	Net efficiency loss
	------- (000 metric tons) -------			(# of workers)	------------------- (million US dollars) -------------------				
Wheat									
Argentina	4,680	660	5,340	78,100	834.5	464.1	204.2	727.0	167.1
Egypt	520	1,820	2,340	88,100	191.6	342.3	-251.9	389.0	101.1
Pakistan	860	1,110	1,970	152,400	304.9	311.4	-39.7	299.7	66.0
Rice									
Argentina	40	10	50	2,100	12.4	8.1	3.2	9.5	0.9
Egypt	2,140	950	3,090	370,200	820.9	393.3	51.4	1,157.5	376.1
Pakistan	930	750	1,680	108,400	629	405.6	107.6	538.1	115.6

Source: Bale and Lutz (1981).

Patterns of Intervention

If one systematically examines the general pattern of price distortions in agriculture across countries there seems to be an association between a country's wealth (per capita income) and the extent to which policies for agriculture move from taxation to protection. This was first noted by Bale and Lutz (1981) and has recently been further studied by Binswanger and Scandizzo (1982). At lower levels of development, countries tend to tax agriculture, resulting in underproduction and overconsumption. At high levels of development, such as in the industrial countries, agriculture is generally subsidized, resulting in overproduction and underconsumption. Countries that lie in between exhibit a mix of disincentives and incentives for agriculture, the mix being dependent on the stage of industrialization that a country has reached and the relative size of agricultural gross domestic product (GDP). These findings are important. As countries graduate to higher levels of development, their agricultural policies become more inward-looking. Agriculture is subsidized and protected from global competition. The "newly industrializing countries" (NICs), taking a leaf from the policy books of industrial countries, are beginning to protect their agricultural sectors. This is done at considerable financial and economic cost to themselves and to their erstwhile trading partners. Thus, while the NICs and developing countries are legitimately speaking out for access to industrial country markets for their manufactures, they are progressively insulating their domestic economies. This brand of "free trade" is called *mercantilism*. International trade is seen as a competitive war that is "won" by those with trade surpluses and "lost" by those with deficits; exports are the objective of trade policy, imports being "the canker at the heart of trade." These views have been proved analytically empty by economists since Adam Smith, but their hold remains unshaken.

Conclusions

What emerges from this discussion is the vital role that agricultural prices play in achieving optimum output, growth, trade, and food security. When governments intervene in the legitimate functioning of markets— even when their actions are well intentioned—the ultimate outcome is to reduce national and global welfare. As individuals with assorted business contacts in developing countries, we should take every opportunity to inform leaders in developing countries of the inherent dangers of market intervention in agriculture.

Notes

1. In making this comparison, farm-gate prices are adjusted upward by transportation costs and marketing margins to "border price." This allows direct comparison with the CIF world price. The comparisons are made for 1976. Further details are found in Bale and Lutz (1981).

References

Bale, M. D. and E. Lutz. "Price Distortions in Agriculture and Their Effects: An International Comparison." *Am. J. Agr. Econ.* 63 (1981): 8–22.

Binswanger, H., and P. L. Scandizzo. "Patterns of Agricultural Protection." World Bank draft working paper, July 19, 1982. Mimeographed.

Lutz, E., and P. L. Scandizzo. "Price Distortions in Developing Countries: A Bias Against Agriculture." *Eur. Rev. Agri. Econ.* 7(1980): 5–27.

Schultz, T. W., ed. *Distortions of Agriculture Incentives.* Bloomington: Indiana University Press, 1978.

Tolley, G. S., V. Thomas, and C. M. Wong. *Agricultural Price Policies and the Developing Countries.* Baltimore: Johns Hopkins University Press, 1982.

World Bank. *Accelerated Development in Sub-Saharan Africa: An Agenda for Action.* Washington, D.C.: World Bank, 1981.

Discussion _____

Robert E. Evenson
Department of Economics
Yale University

The existence of "schools" of thought on a topic generally reveals that the topic is both important and imperfectly understood. Agricultural supply fits into this characterization quite well. Several schools of thought have emerged around the topic, and its importance to human welfare is self-evident. An early school of thought, stressing culture, tradition, and "structural" factors as determinants of supply, has largely been laid to rest by the economic research of the past 25 years or so. However, we still have several recognizable schools with different approaches to the problem of increasing agricultural supply. One of these schools stresses the role of farm prices as guides to efficient resource allocation. One stresses technology production and diffusion. Another stresses rural infrastructure. Still another stresses legal and institutional factors.

In extreme form each of these schools might be termed fundamentalistic. Price fundamentalists, for example, would argue that getting price right not only produces allocative efficiency in a static sense but efficiency in a dynamic sense as well, presumably through induced innovation mechanisms. Technology fundamentalists see the production and diffusion of technology as being only marginally influenced by prices. Infrastructure fundamentalists stress irrigation, roads, and institutions, while market fundamentalists stress contracts, transactions, costs, and legal procedures.

Lucio Reca has taken on the formidable task of summarizing price policies in developing countries. He provides us with a useful classification of objectives, instruments, and effects of price policy. Several specific cases of price policy effects are discussed. It would be unfair to characterize Reca as a price fundamentalist. He does recognize a role for public sector investment in infrastructure and technology discovery. Nonetheless, he does argue for a relationship between long-term supply growth and price regimes that suggests a very strong induced innovation mechanism.

I have no disagreement with the basic comparative static implications of price policies as discussed by Reca. He refers to a significant body of evidence on single crop responsiveness to prices showing that farmers do respond to prices. We have relatively little evidence on many of the cross-price supply effects, however, and very few studies of aggregate price responsiveness. We do have a considerable body of evidence on consumer price responsiveness showing that consumers respond to prices so that a reduction in the price of one commodity will affect the prices of other commodities through the demand side.

There is little doubt that some economies have introduced price policy distortions that have produced large efficiency losses. The EEC countries, Japan, and the United States are notable examples. These, however, are rich countries that can afford the extravagance of these price policies. How about the really poor countries? How large are their stocks of efficiency losses to be reduced by appropriate price policy? Our image of the potential growth to be achieved by getting prices right is heavily conditioned by the distortions of the rich countries. These distortions are not only on a much larger scale than those of the poorer countries; they also run in the opposite direction. Rich countries generally intervene to achieve farm prices above market equilibrium levels. Poor countries generally intervene to achieve farm prices below market equilibrium levels, and in some cases to achieve farm input prices below equilibrium levels.

Professor Reca discusses the objectives of price policy from the perspective of planners and bureaucrats and perhaps of the public relations staff of organized political interest groups. But they are not fair characterizations of the objectives of real interest groups. The wheat producers or the pork producers in a given country are not likely to publicize their central objective, which is to increase the incomes of their group's membership. Instead they will appeal to more general objectives.

Self-sufficiency as an objective is a powerful idea in many countries. Professor Reca does not include it as an objective of price policy in his discussion. When stated as an economic objective, of course, it is flawed because it is inconsistent with the principle of comparative advantage. Nonetheless, it has broad political appeal. Unfortunately, it generally does not produce an improvement in farm output price regimes, but is instead used to justify large project expenditures and price intervention in farm capital inputs.

Reca's discussion of price policy instruments does not fully examine the administrative and enforcement costs of most instruments. In practice, many countries have found these costs to be so high as to severely limit the effectiveness of price policy actions. Indeed, in many countries, the primary interest groups supporting administered prices, import licensing, etc. are the potential rent collectors who engage in bribery and black market activities. The paper presented by Professor Bates (see Chapter 5) documents some of this for Africa. In many cases the administrative and enforcement costs may outweigh the pure efficiency losses from a given price policy by a large margin.

Reca lists administered product prices, administered input prices and subsidies, international trade tariffs and quotas, exchange rate policy, marketing boards, and credit subsidies as the main instruments of price policy. He does not discuss the substitutability between these instruments in terms of achieving a particular objective, but other contributors to this book have pointed out a tendency for poor countries to use a somewhat different mix of instruments than higher-income countries. Schuh (Chapter 8) stresses the effects of overvalued exchange rate policies on the agricultural export sectors of both developed and developing countries. Yet, the primary motivation for overvalued exchange rate policies is probably that they favor certain politically powerful nonagriculture importing sectors. Bates notes the tendency to establish marketing boards in many countries because they can become mechanisms for taxing certain sectors. He also notes their almost universal tendency to overtax and to become costly bureaucracies. Reca points out that virtually every water project ever constructed anywhere in the world underprices water and must be publicly subsidized, attesting further to the importance of the administration and enforcement dimension of price policy instruments.

Reca devotes the major part of his paper to a discussion of some of the effects of price policies. He provides several interesting cases and shows that a price intervention in one commodity can have important effects in other markets because of substitutability and complementarity on both the demand and supply side.

In discussing price policy effects, it is important that a distinction be made between the comparative static effects of a given policy action and the longer-run dynamic aspects. The economic theory that we apply to the analysis and much of the empirical evidence available is suited to the measurement of comparative static effects. (Our empirical evidence, however, is very limited as regards cross-supply effects.) We have little in the way of theory or systematic evidence to address the effect of price regimes on longer-run supply growth possibilities.

The comparative static results discussed by Reca and also summarized by Malcolm Bale are reasonably straightforward. We have a great many estimates of single crop supply responsiveness that can be used to assess the change from one level of supply to another as price levels change. As I noted, we have very few estimates of cross-supply responsiveness. We also have very few "aggregate" supply elasticities measuring the aggregate supply response to a change in all farm output prices relative to farm input prices. The available aggregate supply elasticity estimates are in the 0.2 to 0.4 range. These estimates are not high enough to make the price fundamentalists' case. An increase in farm output prices of 20 percent would induce a supply response of from 4 to 8 percent, but this would be a one-time-only response. Spread over five years it represents a modest growth component. Furthermore, it becomes clear that buying additional growth through further price increases is a very costly way to achieve growth unless the change in the price regime alters the basic dynamics of agricultural growth.

Of course, in this comparative static framework the objective of eliminating a price distortion is not necessarily to increase farm output, but to eliminate inefficiency. Elimination of many price distortions will not produce appreciable output expansion.

A relatively new body of empirical work based on the "duality" relationship between production or transformation functions and maximized profits functions is beginning to produce some additional evidence on cross-supply elasticities and on the effects of output prices on input demands. The methodology allows the specification of a maximized profits function incorporating multiple outputs and multiple variable inputs. The duality theorems allow us to impose conditions on the profits function that will insure that the dual transformation function is well behaved. The first derivative functions of the profits function (Shephard's Lema) with respect to output and input prices then form a system of output supply and input demand functions.

To illustrate the nature of these systems and to provide some added information on price effects I report estimates of elasticities from district level data (1959–1975) for the major wheat producing regions in northern India (Punjab, Haryana, and western Uhar Pradesh). Table 1 reports supply elasticities for four crops or aggregates of crops: wheat, rice, coarse cereals and other crops, estimated using a normalized quadratic functional form. Each supply function is a function of variable output prices, variable input prices, and several "fixed" factors—all presumed to be exogenous to the individual farmer. The "t" statistics reported are asymptotic for the partial derivatives of the system evaluated at the mean levels of the variable. Symmetry restrictions were imposed (see Evenson, 1982, for details).

Table 1 shows that the own supply elasticities for each crop are positive and moderately low. These may be regarded as short-run elasticities, as net cropped area and the proportion irrigated for the

Table 1

North India Wheat--Full System Elasticities

	Wheat	Rice	Coarse Cereals	Other Crops
Price of Wheat	.3180	-.3601	.3227	-.0652
	(3.305)	(2.133)	(2.089)	(1.241)
Price of Rice	-.0706	.3031	-.0742	.0325
	(2.133)	(2.335)	(.8711)	(1.641)
Price of Coarse Cereals	.0867	-.1016	.3507	-.0509
	(2.089)	(.8711)	(2.456)	(1.901)
Price of Other Crops	-.1093	.2780	-.3174	.1613
	(1.241)	(1.641)	(1.901)	(2.141)
Price of Fertilizer	-.0055	.0478	.0016	-.0341
	(.2693)	(.7821)	(.0294)	(2.884)
Price of Bullocks	-.0332	-.1289	-.0483	-.0097
	(1.943)	(2.271)	(.9818)	(1.007)
Price of Tractors	.0027	-.0215	-.0164	.0024
	(.3210)	(.7714)	(.6839)	(.4790)
Wages	-.1888	-.0167	-.2188	-.0364
	(3.037)	(.1577)	(2.148)	(.6921)
Electricity	-.0311	.0039	.0039	.0592
	(1.012)	(.0736)	(.0798)	(2.132)
Roads	-.0138	-.0716	.2460	-.3304
	(.1270)	(.3951)	(1.423)	(3.374)
Rain	.0360	.2124	-.3057	.0835
	(.4603)	(1.609)	(2.450)	(1.177)
High Yielding Varieties	.2096	.4116	-.0339	-.0220
	(7.741	(8.644)	(.8082)	(.8932)
Irrigation	1.4018	.8136	.8725	.7903
	(10.454)	(3.611)	(4.276	(6.481)
Net Cropped Area	-.1411	-.0501	.5418	.3078
	(.4999)	(1.862)	(1.329)	(1.190)

Symmetry Restrictions
F Value: 3.7380
Weighted R-Square .5653
Characteristics Roots of the Hessian Matrix:

Col. 1	Col. 2	Col. 3	Col. 4
634.531	3467.02	1674.45	855.072

Source: Evenson (1982).

farm (not for each crop) is treated as fixed. Note that we obtain significant cross-elasticity effects in some cases. An increase of 10 percent in the price of wheat, holding all other prices and fixed quantities constant, causes a 3.2 percent increase in the supply of wheat, but a 3.6 percent decrease in the supply of rice. Interestingly, it causes a 3.2 percent increase in the supply of coarse cereals (maize, sorghum and millets) because these are summer crop complements to the winter wheat crop. Rice and the other crops (pulses, groundnuts, etc.) form a similarly weak complementary combination where rice is the summer crop.

We may note further that the prices of inputs affect output supply. It is particularly interesting to note that a rise in wages and in bullock prices have the most significant negative effects on supply.

Table 2 presents estimates of an aggregate supply function estimated along with the variable input demands equations. I should note that at the district level we have rather poor input quantity data and these results should be discounted accordingly. (See Evenson and Binswanger, 1979, for estimates of input demand function based on farm management data). Nonetheless, these estimates show that output prices affect aggregate output and the demand for factors. They also show that input prices affect aggregate output.

Price response estimates of this type can enrich our comparative static analysis of price policy effects. They do not, however, speak to the dynamic relationship between prices and agricultural supply. In the estimates summarized in Tables 1 and 2, I have included a variable measuring the availability of high-yielding variety (HYV)—wheat, rice, and maize—technology to the districts. (It is the percentage of total cropped area planted to HYVs). As the estimates show, it has a substantial effect on output growth. This effect is biased on both the output side and the input side. HYVs in the form realized in northern India shift wheat and rice supply functions to the right but shift coarse cereals and other crop supply to the left. They also shift the demand for fertilizer and tractors sharply to the right while having very modest effects on bullock and human labor.

This high-yielding variety effect is not specified in such a way as to adequately test the proposition of an interaction between prices and technology. The fact that we have biases in the technology effect indicates that if prices were changed to favor the biases in technology we should see more rapid adoption and use of the technology. This may be part of what Reca refers to as a "transcendence" relationship (p. 165) between technological and price policy. If prices are rigged against products and factors enjoying favorable biases, this will inhibit growth and efficient use not only of conventional resources but of technology resources as well.

Table 2

North India Wheat--All Crop System Elasticities

	All Crops	Fertilizer	Bullocks	Tractors	Labor
Price of all Crops	.1464 (3.233)	.5164 (3.084)	.1045 (6.174)	.2296 (.9557)	.0039 (.1276)
Price of Fertilizer	-.0195 (3.084)	.0795 (.6977)	-.0460 (4.396)	-.5140 (1.824)	-.2362 (4.703)
Price of Bullocks	-.0298 (6.174)	.3476 (4.396)	-.0304 (1.963)	-.9268 (5.572)	.1632 (5.552)
Price of Tractors	-.0026 (.9557)	-.1532 (1.824)	-.0366 (5.572)	1.2983 (3.753)	-.0909 (1.465)
Wage	-.0946 (2.338)	-.0950 (.8711)	.0085 (.7503)	-.0872 (.5319)	-.0218 (1.011)
Electricity	.0227 (1.1516)	.0402 (.7656)	.0135 (2.564)	-.0974 (1.5406)	-.0045 (.4717)
Roads	.2086 (2.953)	-.5044 (2.738)	-.0960 (4.933)	1.5406 (5.877)	-.3071 (8.668)
Rain	.0448 (.8823)	.5855 (4.442)	-.0002 (.0185)	.4368 (2.379)	.6276 (1.239)
High Yielding Varieties	.0873 (4.913)	.6058 (12.620)	.0272 (5.456)	.3436 (5.028)	.0919 (10.221)
Irrigation	1.0438 (11.859)	2.827 (12.002)	-.0693 (2.752)	3.958 (11.551)	.1739 (3.814)
Net Cropped Area	.0948 (.5035)	-.7164 (1.492)	-.1928 (3.571)	-3.132 (4.398)	.2500 (2.630)

Symmetry Restrictions
F value: 6.1345
Weighted R-square: .6116
Characteristic Roots of the Hessian Matrix:

Col. 1	Col. 2	Col. 3	Col. 4
7819.3	1218.7	-1610.1	-2166

Source: Evenson (1982).

The technology diffusion literature supports the contention that prices guide rates of adaption of technology. Do they also guide the development of technology? The induced innovation literature provides a theory and some evidence regarding the role of prices in determining paths or biases in technology production, given an existing innovation potential and a willingness to invest in its discovery. This literature has little to say regarding the incentives and motives to invest in the creation of invention potential or the exploitation of that potential, particularly when public sector institutions are involved. Indeed, the large public sector role in the development of agricultural technology is due to a total lack of private sector incentives to invest in research. It is not clear whether price regimes more favorable to farm suppliers would materially alter this incentive situation. The private sector is not a significant source of improved agricultural technology in developing countries today. With higher farm prices we would probably see a growth in the sale of farm inputs to the farming sector, and this would provide incentives for expanded private sector activity. Yet, in low wage economies this purchased input sector will not be very large.

Public sector research institutions almost certainly respond to prices. When the real price of fertilizer falls, the value of fertilizer responsiveness in plants rises and well-managed public research stations will alter plant breeding strategies accordingly.

Public research stations could in principle also act to "correct" for price policy distortions. They might, for example, be guided by "border" prices if influenced sufficiently by economists. Research programs might also be pushed by governments as a substitute for price incentives— they are probably considered by many countries to be part of the "project" strategy to increase farm output. It would appear, however, that factors that affect adoption of technology are probably also affecting research and invention strategies. One bit of evidence for this is available from Argentina. An unpublished study by Evenson and Cordomi (1969) reported that from 1945 to 1956, prices paid to sugarcane producers were based almost entirely on tonnage of cane delivered to the mills. The sucrose content of the cane was ignored as a factor in the pricing of cane. In response to this policy, the average kilograms of sucrose per m. ton of cane declined from 0.0740 in 1944 (0.0780 in 1928) to 0.0674 in 1953. In 1964, when sucrose content was again a factor in cane pricing, it had risen to 0.0827. An examination of varietal adaptions showed that Argentine producers had adopted two varieties specifically rejected as commercial varieties (not recommended to growers) because of low sucrose content in other countries. These two varieties, C0421 from India and CP29-116 from Florida, and local variety TUC2645 (also noted for low sucrose content) achieved significant adoption in

Argentina during the period. These varieties were adapted instead of several economically superior varieties produced by the Tucuman station in the late 1930s and 1940s. The inefficient price signals appeared to cause a divergence of research effort toward high cane yielding, low sucrose content varieties in the 1950s in Tucuman. This research effort was rendered useless by the return to sucrose content pricing in the 1960s.

It would appear that the simple comparative static analyses of price policy effects do not give the full story. Price regimes do affect the adaption and diffusion of technology. They have a role in inducing the development of technology as well. Simply "getting prices right," however, will not insure optimal agricultural growth. The public sector has a major role to play and this role is not necessarily complementary to price policy actions.

References

Evenson, R. E., and H. Binswanger. "The Demand for Labor in Indian Agriculture." In H. Binswanger and Rosenzwing, *Rural Labour Markets in Asia*. New Haven, Conn.: Yale University Press, 1983.

Evenson, R. E., and M. Cordomi. "Responsiveness to Economic Incentives by Sugarcane Producers in Tucuman, Argentina." Economic Growth Center, Yale University, 1970. Mimeographed.

Evenson, R. E. "The Green Revolution in North India." Economic Growth Center, Yale University, New Haven, Conn., 1982. Mimeographed.

Part 4 _____

5

Governments and Agricultural Markets in Africa

Robert H. Bates
California Institute of Technology
Pasadena, California

Governments in Africa intervene in agricultural markets in characteristic ways. They tend to lower the prices offered for agricultural commodities. They tend to increase the prices which farmers must pay for the goods they buy for consumption purposes. And although African governments subsidize the prices that farmers pay for the goods they use in farming, the benefits of these subsidies are appropriated by the rich few—the small minority of large-scale farmers.

There are other characteristics of patterns of government market intervention. Insofar as African governments seek increased farm production, their policies are project-based rather than price-based. Insofar as they employ prices to strengthen production incentives, they tend to encourage production by lowering the prices of inputs (i.e., by lowering costs) rather than by increasing the prices of products (i.e, by increasing revenues). A last characteristic is that governments intervene in ways that promote inefficiency; they create major price distortions, reduce competition in markets, and invest in poorly conceived agricultural projects. In all of these behaviors, it should be stressed, the conduct of African governments resembles the conduct of governments in other parts of the developing world.

One purpose of this paper is to describe more fully these patterns of government intervention. A second is to examine a variety of explanations for this behavior.

The Regulation of Commodity Markets[1]

It is useful to distinguish between two kinds of agricultural commodities: food crops, many of which could be directly consumed on the farm, and cash crops, few of which are directly consumable and which are instead marketed as a source of cash income. Many cash crops are in fact exported; they provide not only a source of cash incomes for farm families but also a source of foreign exchange for the national economies of Africa.

Export Crops

An important feature of the African economies is the nature of the marketing systems employed for the purchase and exportation of cash crops. The crops are grown by private farm families, but they are then sold through official, state-controlled marketing channels. At the local level, these channels may take the form of licensed agents or registered private buyers; they may also take the form of cooperative societies or farmers' associations. But the regulated nature of the marketing system is clearly revealed in the fact that these primary purchasing agencies can in most cases sell to but one purchaser: a state-owned body, commonly known as a marketing board.

Marketing Boards

Background. The origins of these boards are diverse. In some cases, particularly in the former settler territories, they were formed by farmers themselves. At the time of the great depression, commercial farmers banded together in efforts to "stabilize" the markets for cash crops; in effect, with the support of the colonial states they dominated, they sought to create producer-dominated cartels. More commonly, the origins of the marketing boards lay in an alternative source of cartel formation: in the efforts of the purchasers and exporters of cash crops to dominate the market and to force lower prices on farmers (Bauer 1964; Jones 1980).

In either case, it was the second world war which led to the institutionalization of the regulation of export markets. During the war, Britain sought to procure agricultural commodities and raw materials from her colonial dependencies. Some materials, such as food for troops in North Africa, were needed for the war effort; others were needed to generate foreign exchange for the purchase of armaments from North America; and the purchase of still other goods was required to provide prosperity for the colonial areas and thereby to lessen the likelihood of political instability at a time in which British armed forces were already spread perilously thin. To secure the regularized purchase of raw materials, the British government created a Ministry of Supply. The ministry signed bulk purchasing agreements with the colonial governments in each of the African territories. To administer the terms of these agreements, the colonial authorities created official state marketing agencies. In those territories in which large-scale producers had already begun to operate "market stabilizing" schemes, the organizations running these schemes were essentially recruited to staff and administer the state marketing boards. In the territories where purchasers' cartels held a predominance of market power, the state procurement schemes essentially gave a legal framework for the merchant-based cartels; the cartels became the instruments for securing raw materials (Leubuscher 1956; Huxley 1957, pp. 137ff; Bauer 1964).

In either case, upon independence, many African governments found themselves the inheritors of bureaucracies that held a legal monopoly over the purchase and export of commodities in the most valuable sector of their domestic economies. These new states possessed extremely powerful instruments of market intervention. They could purchase export crops at an administratively set, low domestic price; they could then market these crops at the prevailing world price; and they could accumulate the revenues generated by the difference between the domestic and world prices for these commodities.

Government Taxation. Initially, the revenues accumulated by the marketing boards were to be kept in the form of price assistance funds and used for the benefit of the farmers. At times of low international prices, they were to be employed to support domestic prices and so shelter the farmers from the vagaries of the world market. In the case of the Western Nigerian marketing board, for example, 70 percent of the board's revenues were to be retained for such purposes. But commitments to employ the funds for the benefit of the farmers proved short-lived. They were overborn by ambitions to implement development programs and by political pressures brought to bear upon governments from nonagricultural sectors of the economy.

One example is the Cotton Price Assistance Fund, accumulated by the Lint Marketing Board in Uganda. Employed to stabilize prices in the 1950s, it was increasingly used thereafter for other purposes. In the pre-Independence period, for example, it was used to secure revenues for the building of the Owen's Falls Dam; although the fund purchased shares in the Uganda Electricity Board—the agency responsible for the dam—it has received no dividends from these shares (and they have declined in value). In the 1960s, the fund "loaned" 100 million Ushs. to the government for investment in the capital budget, interest free! Still later, it was employed to capitalize the Cooperative Development

Bank with a 12 million Ushs. contribution, again interest free, repayable over 35 years (Uganda 1977; Walker and Ehrlich 1959).

Similarly, in West Africa, the revenues of the marketing boards were increasingly diverted to uses other than the stabilization of farmers' incomes. In Nigeria, for example, funds were first loaned to the regional governments; later, they were given to these governments in the form of grants; later still, the legislation governing the use of these revenues was altered such that the boards became instruments of direct taxation (Onitiri and Olatunbosun 1974). We have already noted that the statutes governing the marketing boards in Western Nigeria reserved 70 percent of the trading surpluses for price stabilization; an additional 7.5 percent was to be employed for agricultural research and the remaining 22.5 percent for general development purposes. But Helleiner (1960) notes that following self-government; "The Western Region's 1955–1960 development plan announced . . . the abandonment of the '70-22.5-7.5' formula for distribution of the Western Board's right to contribute to development, and provided for £20 million in loans and grants to come from the Board for the use of the Regional Government during the plan." The board "was now obviously intended to run a trading surplus to finance the regional Government's program. The Western Region Marketing Board had by now become . . . a fiscal arm of the Western Nigerian Government" (pp. 170–171). This transition was followed as well in Ghana, where "the government decided to remove . . . legal restrictions on its access to the funds of the Board" (Beckman 1976, p. 199).

The movement from an instrument of price stabilization, largely for the benefit of farmers, to an instrument of taxation, with the diversion of revenues to nonfarm sectors, can be seen as well in changes in the pricing formulas employed by the marketing boards. Insofar as the boards were employed to stabilize producer prices, the domestic prices— i.e., the price offered the farmers—should have moved independently of the world prices; moreover, a policy of price stabilization implies that domestic prices should have at times exceeded world prices, as the marketing board attempted to protect farmers from falls in the world price. But investigations clearly suggest that what was being stabilized was not the domestic price but rather the difference between the domestic and world price, i.e. the tax on the farmers' income (see Bates 1981).

Food Crops

African governments also intervene in the market for food crops. And, once again, they tend to do so in ways that lower the price of agricultural commodities.

Table 1

Patterns of Market Intervention for Food Crops

Crop	Countries in Which Crop Is Grown	Countries with Producer price Controls		Countries with Legal Monopoly Over Crop	
	N	N	%	N	%
Rice	26	25	96	11	42
Wheat	12	8	67	4	33
Millet and Sorghum	38	9	24	7	18
Maize	35	24	69	9	26
Roots and Tubers	33	6	18	1	3

<u>Source</u>: United States Department of Agriculture. <u>Food Problems and</u>

<u>Prospects in Sub-Saharan Africa</u> (Washington, D.C.: USDA.

1980), p. 173.

Price Controls. One way in which African governments attempt to secure low-priced food is by constructing bureaucracies to purchase food crops at government-mandated prices. A recent study by the U.S. Department of Agriculture examined the marketing systems for food crops in Africa and discovered a high incidence of government market intervention. In the case of three of the food crops studied, in over 50 percent of the countries in which the crop was grown, the government had imposed a system of producer price controls; in over 20 percent the government maintained an official monopsony for the purchase of that food crop (Table 1).

The regulation of food markets entails policing the purchase and movement of food stocks and control over the storage, processing, and retail marketing of food. An illustration is offered by the maize industry of Kenya; according to subsection 1 of section 15 of the Maize Marketing Act (Schmidt 1979, p. 25):

All maize grown in Kenya shall, subject to the provision of this Act, be purchased by and sold to the Board, and shall, without prejudice to the Board's liability for the price payable in accordance with section 18 of this Act, rest in the Board as soon as it has been harvested.

According to the Maize Marketing (Movement of Maize and Maize Products) Order, all movements of maize require a movement permit, valid for only 24 hours, which must be obtained from the Maize and Produce Board. The sole exceptions are the movement of maize or maize products within the boundaries of the farm, the movement of not more than two bags (180 kg) accompanied by the owner, and the movement of not more than ten bags within the boundaries of a district accompanied by the owner and intended for consumption by the owner or his family (Schmidt 1979).

The impact of controls over the market for food crops is profound. The costs of marketing increase. In part, this is because the government marketing board is less efficient than the private sector in the transport and storage of maize; and in part it is simply because the government-imposed barriers to entry in the maize market confer excess profits on the agents who remain within the market. The nature and the magnitude of these higher costs is perhaps most vividly illustrated in the "bribe costs" that those operating in the regulated market can impose. According to Schmidt (1979, p. 68), "Bribing costs were not simply a problem with regard to illegal movements of maize and beans. More than 90 percent of the . . . agents mentioned this . . . in regard to deliveries to [Maize and Produce Board] depots. In fact, in some areas the problem was so severe that bribes were the major cost item for agents. Bribing is sometimes necessary for virtually all steps to get maize into the depots: obtaining movement permits, passing the gate, passing the moisture test, getting the lorry off-loaded and so forth."

A second major consequence of the regulated maize market is price inefficiency. Under the present system, interregional price differentials exceed interregional costs of transport and intertemporal price differentials exceed the costs of storage. The result is that many consumers pay higher prices and many producers receive lower prices than would be the case were maize more easily moved between places and over time. With a more efficient marketing system, farmers in places or periods of surplus could more easily consummate deals with consumers in places or periods of food deficit, deals from which both parties could reap an advantage. These unconsummated transactions constitute a loss of economic welfare.

More directly relevant to the concerns of this paper, however, is the impact of the food marketing controls on food prices. For insight into this subject we can turn to Doris Jansen Dodge's (1977) study of NAMBoard, the food marketing bureaucracy in Zambia. Over the years studied by Dodge (1966/1967 to 1974/1975) NAMBoard depressed the price of maize as much as by 85 percent; in the absence of government controls over maize movements, the farmers could have got up to 85 percent more for their maize than they were able to secure under the market controls imposed by NAMBoard. Gerrard (1981) extends Dodge's finding for Zambia to Kenya, Tanzania and Malawi; Dodge herself extends them to eight other African countries (see Jansen 1980).

Projects. In order to keep food prices low, governments take additional measures. In particular, they attempt to increase food supplies, either by importing food or by investing in food production projects.

Foreign exchange is scarce. Especially since the rise of petroleum prices, the cost of imports is high. To conserve foreign exchange, African governments therefore attempt to become self-sufficient in food. But they seek to do so within the context of a low-price policy; and therefore invest in projects that will yield increased food production.

In some cases, governments turn public institutions into food production units: youth league and prison farms provide illustrative cases. In other instances, they invest in large-scale efforts to furnish scarce factors of production. In Africa, water is commonly scarce and governments invest heavily in river basin development schemes and irrigation projects. Capital equipment is also scarce; by purchasing and operating farm machinery, governments attempt to promote farm production. Some governments invest in projects to provide particular crops: rice production in the case of Kenya, for example, or wheat production in the case of Tanzania. In other instances, governments divert large portions of their capital budgets to the financing of food production schemes. Western Nigeria, for example, spent over 50 percent of the Ministry of Agriculture's capital budget on state farms over the period of the 1962–1968 development program (Hill 1977; Roider 1971).

Nonbureaucratic Forms of Intervention

Thus far I have emphasized direct forms of government intervention. But there is an equally important, less direct form of intervention: the overvaluation of the domestic currency.

Most governments in Africa maintain an overvalued currency (IBRD 1981; Picks 1978). One result is to lower the prices received by the exporters of cash crops. For a given dollar earned abroad, the exporters of cash crops receive fewer units of the domestic currency. In part, overvaluation also inflicts losses on governments; deriving a portion of their revenues from taxes levied by the marketing boards, the governments command less domestic purchasing power as a result of overvaluation. But because their instruments of taxation are monopolistic agencies, African governments are able to transfer much of the burden of overvaluation: they pass it on to farmers in the form of lower prices.

In addition to lowering the earnings of export agriculture, overvaluation lowers the prices paid for foreign imports. This is, of course, part of the rationale for a policy of overvaluation: it cheapens the costs of importing plant, machinery and other capital equipment needed to build the base for a nascent industrial sector. But things other than plant and equipment can be imported, and among these other commodities is food. As a consequence of overvaluation, African food producers face higher levels of competition from foreign food stuffs. In search of low-priced food, African governments do little to protect their domestic food markets from foreign products—whose prices have artificially been lowered as a consequence of public policies.

Industrial Goods

In the markets for the crops they produce, African farmers therefore face a variety of government policies that serve to lower farm prices. In the markets for the goods they consume, however, they face a highly contrasting situation: they confront consumer prices that are supported by government policy.

In promoting industrial development, African governments adopt commercial policies that shelter local industries from foreign competition. To some degree, they impose tariff barriers between the local and international markets. To an even greater extent, they employ quantitative restrictions. Quotas, import licenses, and permits to acquire and use foreign exchange: all are employed to conserve foreign exchange on the one hand, while on the other hand they protect the domestic market for local industries. In connection with the maintenance of overvalued currencies, these trade barriers create incentives for investors to import capital equipment and to manufacture goods domestically that formerly had been imported from abroad (Stryker 1975; Pearson et al. 1981; IBRD 1975, 1978).

Not only do government policies shelter industries from low-cost foreign competition; they shelter them from domestic competition as well. In part, protection from domestic competition is a by-product of protection from foreign competition. The policy of allocating licenses to import in conformity with historic market shares provides an example. The limitation of competition results from other policies as well. In exchange for commitments to invest, governments guarantee periods of freedom from competition. Moreover, governments tend to favor larger projects; seeking infusions of scarce capital, they tend to back those proposals that promise the largest capital investments. Given the small markets typical of most African nations, the result is that investors create plants whose output represents a very large fraction of the domestic market; a small number of firms thus come to dominate the industry. Lastly, particularly where state enterprises are concerned, governments sometimes confer virtual monopoly rights upon particular enterprises. The consequence of all these measures is to shelter industries from domestic competition.

One result is that inefficient firms survive. Estimates of the use of industrial capacity range as low as one-fifth the single shift capacity of installed plants (state enterprises in Ghana in 1966, Killick 1978, p. 171). Another consequence is that prices rise. Protected from foreign competition and operating in oligopolistic or monopolistic settings, firms are able to charge prices that enable them to survive despite very high operating costs.

Farm Inputs

By depressing the prices offered farmers for the goods they sell, government policies lower the revenues of farmers. By raising the prices consumers—including farmers—must pay, governments reduce the real value of farm revenues still further. As a consequence of these interventions by governments, then, African farmers are taxed. Oddly enough, while taxing farmers in the market for products, governments subsidize them in the market for farm outputs.

Attempts to lower input prices take various forms. Governments provide subsidies for seeds and fertilizers, with fertilizer subsidies running from 30 percent in Kenya to 80 percent in Nigeria. They provide tractor hire services at subsidized rates, up to 50 percent of the real costs in Ghana in the mid-1970s (Stryker 1975; Kline et al. 1967). They grant loans at subsidized rates of interest for the purchase and rental of inputs, and they provide highly favorable tax treatment for major investors in commercial farming ventures (see, for example, Ekhomu 1978). Moreover, through their power over property rights, African governments have released increased amounts of land and water to commercial farmers at costs that lie below the value they would generate in alternative uses. The diversion of land to large-scale farmers and of water to private tenants on government irrigation schemes—without compensating those who employed these resources in subsistence farming, pastoral production, fishing or other ventures—represents the conferring of a subsidy upon the commercial farmer at the expense of the small-scale, traditional producer. This process has been documented in northern Ghana (USAID 1975; *West Africa*, 3 April 1978); Nigeria (Ekhomu 1978; Girdner and Olorunsula 1978); Kenya (Njonjo 1977); Ethiopia (Cohen and Weintraub 1975) and Senegal (Cruise O'Brien 1971). It was, of course, common in settler Africa as well.

In the case of land and water, then, a major effect of government intervention in the market for inputs has been to augment the fortunes of large-scale farmers at the expense of small-scale farmers. To some degree, this is also true of programs that support chemical and mechanized inputs. Even where there is no direct redistribution, however, it is clear that government programs that seek to increase food production by reducing the costs of farming reach but a small segment of the farming population: the large farmers. In part, this is by plan: the programs are aimed at the "progressive farmers" who will "make best use of them." In part, it is because the large farmers share a common social background with those who staff the public services; the public servants therefore aim their programs and services at those with whom they feel they can work most congenially and productively (see Leonard 1977, and Van Velsen 1973). And, in part, it is because the favoring of the large farmer is politically productive. I will elaborate upon this argument below.

Discussion

Governments intervene in the market for products in an effort to lower prices. They adopt policies that tend to raise the price of the goods farmers buy. And while they attempt to lower the costs of farm inputs, the benefits of this policy are experienced by a small minority of the richer farmers. Agricultural policies in Africa thus tend to be adverse to the interests of most producers.

Studies in other areas suggest that this configuration of pricing decisions is common in the developing nations (Krishna 1967; U.S. GAO 1975; Gotsch and Brown 1980; Griffin 1972; Lipton 1977). Indeed, it is argued by some that the principal problems bedeviling agriculture in the developing areas originate from bad public policies. In the words of Schultz, given the right incentives, farmers in the developing world would "turn sand into gold" (Schultz 1976, p. 5). Distortions introduced into agricultural markets by governments, he contends, furnish the most important reasons for their failure to do so (Schultz 1978). Although Schultz's position is perhaps an extreme one, it nonetheless underscores the importance of understanding why Third World governments select this characteristic pattern of agricultural policies. In the sections remaining, I will advance several explanations for their choices.

Governments as Agents of the Public Interest

The first approach, which derives from development economics, regards governments as agencies whose task is to secure the best interests of their societies. Public policy represents choices by the government made out of a regard for what is socially best. The overriding public interest of poor societies is rapid economic growth, and the policy choices of Third World governments represent their commitment to rapid development, a commitment that implies supplanting agriculture with industry.

In common with most political scientists, I regard the political theory of welfare economics—of which development economics is but that branch applied to poor societies—as being invariably naive, often wrong, and occasionally pernicious. It is therefore unsettling to have to admit that, *confining our attention to export crops*, the implications of this approach are consistent with many of the facts.

All the governments of Africa seek industrial development. Most seek to create the social and economic infrastructure necessary for industrial growth, and many are committed to the completion of major industrial and manufacturing projects. To fulfill their plans, governments need revenue; they also need foreign exchange. In most of the African nations, agriculture represents the single largest sector in the domestic economy; and in many it represents the principle source of foreign exchange. It is therefore natural that in seeking to fulfill their objectives for their societies, the governments of Africa should intervene in markets in an effort to set prices in a way that transfers resources from agriculture to the "industrializing" sectors of the economy: the state itself and the urban industrial and manufacturing firms.

An explanation based on the development objectives of African regimes is thus consistent with the choices made in the markets for export goods. It is also consistent with other well-known facts. The policy choices which have been made are, for example, in keeping with the prescriptions propounded in leading development theories. According to these theories, to secure higher levels of per capita income, nations should move from the production of primary products to the productions of manufactured goods. Savings take place out of the profits of industry and not out of the earnings of farmers. Resources should therefore be levied from agriculture and channeled into industrial development. And agriculture in the developing areas, it is held, can surrender revenues without significant declines in production. These were, and remain today, critical assertions in development economics. Many policymakers in Africa were trained under development specialists; and important advocates of these arguments have served as consultants to the development ministries of the new African states. It is therefore credible to account for the policy choices made by African governments—which systematically bias the structure of prices against agriculture and in favor of industry—as choices made in accordance with prescriptions of how best to secure the welfare of people in poor societies.

Such an approach ultimately proves unsatisfactory, however, and for several reasons. One set of reasons accepts the legitimacy of this form of explanation; the other does not.

Internal Critique

We may accept for the moment the premise that states act as agencies for maximizing the social welfare. Nonetheless, we are left with the fact that this premise is not very useful, particularly when applied to food policy, for it yields little by way of predictive power. To secure social objectives, governments can choose among a wide variety of policy instruments; and knowledge of the public objectives of a program often does not allow us to predict or to explain the particular policy instrument chosen to implement it.

For example, an important objective of African governments is to increase food supplies. To secure greater supplies, governments could offer higher prices for food or invest the same amount of resources in food production projects. There is every reason to believe that the former is a more efficient way of securing the objective. But governments in Africa systematically prefer project-based policies to price-based policies.

To strengthen the incentives for food production, African governments can increase the price of farm products or subsidize the costs of farm implements. Either would result in higher profits for producers. But governments prefer the latter policy.

To increase output, African governments finance food production programs. But given the level of resources devoted to these programs, they often create too many projects; the programs then fail because resources have been spread too thin. Such behavior is nonsensical, given the social objectives of the program.

To take a last example: In the face of shortages, governments can either allow prices to rise or they can maintain lower prices while imposing quotas. In a variety of markets of significance to agricultural producers African governments choose to ration. They exhibit a systematic preference for the use of this technique—a preference that cannot readily be accounted for in terms of their development objectives.

A major problem with an approach that tries to explain agricultural policies in terms of the social objectives of governments, then, is that the social objectives underlying a policy program rarely determine the particular form the policies assume. The approach thus yields little predictive power. There is a second major difficulty. Insofar as this approach does make predictions, they are often wrong.

This problem is disclosed by the self-defeating nature of many government policies. To secure cheaper food, for example, governments lower prices to producers; but this only creates shortages which lead to *higher* food prices. To increase resources with which to finance programs of development, governments increase agricultural taxes; but this leads to declines in production and to shortfalls in public finances and foreign exchange. And to secure rapid development, governments seek to transfer resources from agriculture to industry; but this set of policies has instead led to reduced rates of growth and to economic stagnation.

The policy instruments chosen to secure social objectives are thus often inconsistent with the attainment of these objectives. The approach thus makes false predictions and it should therefore be rejected. And yet the choices of governments are clearly stable; despite undermining their own goals, governments continue to employ these policy instruments. Some kind of explanation is required, and other kinds of theories must therefore be explored.

External Critique

There are other grounds for rejecting the development economics approach to the explanation of governmental behavior. One is that the approach assumes autonomy on the part of governments: they are viewed as having the capability of making meaningful choices. It could be that domestic forces impose binding constraints on governments in the developing areas; alternatively, their position in the international political economy may offer them a highly impoverished menu of alternatives. In either case, it would make little sense to view governments as possessing the capacity for making choices. Another basis for rejecting this approach is that it posits benign motives for governments. In contrast to welfare economists, political scientists like myself view governments as possessing their own private agendas, and regard it as the duty of all who bear a commitment to the public interest to make it in the private interest of governments to do the same. Quite apart from philosophic predisposition, however, recent experiences in Africa and elsewhere make it clear that the preference of governments often bears little correspondence to any idealization of the public interest. Rather, governments engage in bureaucratic accumulation and act so as to enhance the wealth and power of those who derive their incomes from the public sector; they also act on behalf of private factions, be they social classes, military cliques, or ethnic groups. They engage in economic redistribution, often from the poor to the rich and at the expense of economic growth. These are central themes in policy formation in Africa and their prominence serves to discredit any approach based on a conviction that governments are agencies of the public interest.

Governments as Agents of Private Interests

In the sections which follow, I explore two approaches which reject the assumptions that governments can exercise discretion and that they act out of a regard for the social welfare. Other than joining in this common dissent, however, these approaches possess little in common; for one derives from contemporary Marxism and the other from the liberal theory of pluralism, Marxism's basic ideological rival.

Dependency Theory

Over the last decade, the dominant approach to development studies could loosely be termed "dependency theory."[2] Although many significant issues divide the proponents of this approach, its advocates tend to concur in their subscription to several basic postulates. The most fundamental is that the relevant level of analysis is "macro" in the broadest sense: there is but one world economy, it is capitalist, and the performance of national economies is determined by their location within the world capitalist system. In conformity with the law of uneven development in capitalist societies, some nations—almost all lying in the Northern Hemisphere—form the core of this system; the remainder belong to the periphery. The superior performance of the core economies results from the dynamics unleashed by the accumulation of capital—an accumulation that has in significant part resulted from the expropriation of economic surplus from the underdeveloped nations. The underdevelopment of the less-developed countries is thus not a prelude to their subsequent development; rather, the peripheral nations remain underdeveloped precisely because they bear and have borne much of the cost of developing the capitalist world.

Additional assumptions pertain to the role of government and of agriculture. In certain respects, governments are viewed as irrelevant; given the overpowering importance of structural location within the world economy as a determinant of economic performance, domestic policy choices assume little significance. As one of the major spokesmen for this approach has argued, an ironic corollary for this radical line of analysis is the ultimate irrelevance of ideology: whether the peripheral countries adopt socialism, communism, or capitalism makes little difference, given the limits placed upon their economic performance by their location in the world economy (Wallerstein 1974a, 1974b). More commonly, however, dependency theorists stress that governments in the periphery are "weak" and that they lack autonomy. By this is meant that these governments are unable to defend local interests because political and economic ties bind them to the interests of the capitalist core; they perpetuate the dominance of the core by facilitating the expropriation of surplus from the local economies.

Agriculture occupies a critical place in this theory, for within the world capitalist system, the peripheral nations are held to specialize in the production of primary products. Exporting raw materials, the developing economies face long-term declines in the terms of trade between primary and manufactured goods. This secular trend transforms their economic position within the world system so that greater quantities of resources must be exchanged to secure a constant quantity of imports. As a consequence, agriculture offers opportunities for the transfer of economic surplus. Because the export sector is the leading sector in the peripheral economies, governments systematically favor it. Indeed, the public sector is often dominated by the needs of the export enclave; agrarian "feudal elites" who produce export crops possess disproportionate influence because of their key importance in the foreign trade. In addition to providing the export sector with public services and infrastructure, governments also attempt to supply a pool of cheap labor. They systematically discriminate against the peasant producers of food crops so as to produce cheap food for plantation workers and to increase the supply of labor. Working in the interests of the developed nations, the governments of the peripheral nations thus systematically favor export agriculture, and facilitate the provision of a stream of raw materials for processing in the industrial and manufacturing centers of the core.

International trade barriers, maintained by developed nations, prevent the LDCs from producing processed goods for export. As for manufacturing for the local market, this too is policed in the interests of the core economies. Local manufacturing is often dominated by foreign capital. Through patents and licensing agreements, the manipulation of accounting procedures, invoicing, and other dodges, multinational corporations are able to export the profits they earn locally to the core economies. Possessing a weak industrial base and forced to rely on raw material exports for their economic well-being, the peripheral economies thus fail to accumulate surpluses locally, and the structure of the world capitalist system, marked by the split between the core and periphery, is reproduced through time.

As a synopsis, this account fails to do justice to this form of analysis. It does serve to indicate the theory's conviction in the limited capacity for policy choice available to peripheral governments and its denial of their capacity, much less their desire, to serve public interests. The account also serves to indicate the failings of this approach, which can be briefly summarized.

A clear implication of this theory is that export agriculture is maintained as a privileged enclave in the developing areas. In fact, historically, export agriculture in Africa has been favored; food crops destined for the local market have rarely received equal levels of public investment. And yet it is not true that the primary beneficiaries have always been foreign interests; particularly in more recent times, as we have noted, the proceeds of this industry have been expropriated *locally*. Moreover, it is clear that many of the funds seized from export agriculture are channeled to local manufacturing interests who aspire to rival international capital.[3] Rather than standing as a privileged sector, then, export agriculture often stands as a target of political action, one that is employed to accumulate capital locally rather than to facilitate the growth of the developed economies.

A related failure of this approach is its assumption of the infeasibility of industrial growth in the economies of the less-developed nations. The experiences of non-African economies suggest that manufacturing can—and has—developed rapidly in the "peripheral" economies. And, as we have seen, African governments seek to devise economic environments in which industry can prosper, even, if needs be, at the expense of agriculture. The cases of Nigeria, Kenya, and the Ivory Coast suggest that these efforts can succeed; industrialization lies within the reach of some African economies. Even more to the point, these cases suggest the extent to which such industrialization can be rendered "indigenous," leading to the formation of a stratum of local entrepreneurs and capitalists whose interests lie in local accumulation and investment rather than in the service of profit centers located abroad (see Schatz 1979; Sklar 1979; and Swainson 1977). Industrialization need not represent the triumph of international capital.

Also relevant to this issue is the pattern of government intervention in food markets. Rarely in Africa are the primary beneficiaries of cheap food policies plantations producing export agriculture; rather, the primary beneficiaries tend to be domestic manufacturing firms. The imperative of domestic accumulation thus shapes important characteristics of food policy as well.

The criticisms offered thus far find a common source in a third major failing of this approach: the failure to appreciate the power of governments. It is the governments of Africa that have led the way in restructuring export markets to accumulate capital from agriculture. It is the governments that, through programs of import substitution, have built up local manufacturing. It is the governments that have seen to it that the benefits of their programs have increasingly been channeled to the local bourgeoisie, whether through credit programs, development programs, the regulation of shareholding and financing, or the passage of laws governing the composition of directorships and the nationality of management. Moreover, it is the action of governments in the developed world that has generated political conflicts within the core, and the actions of governments in the periphery have enabled the developing nations to exploit these conflicts to their advantage. These governmental actions underscore the futility of conceiving of the world as being composed of a united and homogeneous core locked in conflict with a passive and exploited periphery.[4]

Pluralist Theory

There exists another alternative. This pluralist approach[5] views public policy as the outcome of political pressures exerted by members of the domestic economy, i.e., by local groups seeking the satisfaction of their private interests from political action.

Particularly in the area of food price policy, this approach has much to recommend it. Put bluntly, food policy in Africa appears to represent a form of political settlement designed to bring peaceful relations between African governments and their urban constituents. It is a settlement whose costs tend to be borne by the farmers.

The urban origins of African food policies are perhaps most clearly seen in Nigeria. If one looks at the historical origins of government food policy in Nigeria, one is drawn to the recommendations of a series of government commissions—the Udoji Commission, the Adebo Commission, and the Anti-Inflation Task Forces, for example—impaneled to investigate sources of labor unrest and to resolve major labor stoppages (see Nigeria 1962, 1974, 1975). The fundamental issue driving urban unrest, they noted, was concern with the real value of urban incomes and the erosion of purchasing power because of inflation. While recommending higher wages, these commissions also noted that pay increases represented only a short-run solution; in the words of the Adebo Commission, "It was clear to us that, unless certain recommended steps were taken and actively pursued, a pay award would have little or no meaning." "Hence," the commission added, "our extraordinary preoccupation with the causes of the cost of living situation" (Nigeria 1975, p. 10). As part of its efforts to confront the cause of the rising cost of living, the commission went on to recommend a number of basic measures, among them proposals "to improve the food supply situation" (p. 93). The origins of many elements of Nigeria's agricultural program lie in the recommendations of these reports.

Urban consumers in Africa constitute a vigilant and potent pressure group demanding low-priced food. Because they are poor, they spend much of their income on food; most studies suggest that urban consumers in Africa spend between 50 and 60 percent of their incomes on food (Kaneda and Johnston 1961). Changes in the price of food therefore have a major impact on the economic well-being of urban dwellers in Africa, and this group pays close attention to the issue of food prices.

Urban consumers are potent because they are geographically concentrated and strategically located. Because of their geographic concentration, they can quickly be organized; because they control such basic services as transport, communications, and public services, they can impose deprivations on others. They are therefore influential. Urban unrest forms a significant prelude to changes of governments in Africa, and the cost and availability of food supplies is a significant factor promoting urban unrest.

It is not only the worker who cares about food prices. Employers care about food prices because food is a wages good; with higher food prices, wages rise and, all else being equal, profits fall. Governments care about food prices not only because they are employers in their own right but also because as owners of industries and promoters of industrial development programs they seek to protect industrial profits. Indicative of the significance of these interests is that the unit that sets agricultural prices often resides not in the Ministry of Agriculture but in the Ministry of Commerce or Finance.

When urban unrest begins among food consumers, political discontent often rapidly spreads to upper echelons of the polity: to those whose incomes come from profits, not wages, and to those in charge of major bureaucracies. Political regimes that are unable to supply low-cost food are seen as dangerously incompetent and as failing to protect the interests of key elements of the social order. In alliance with the urban masses, influential elites are likely to shift their political loyalties and to replace those in power. Thus it was that protests over food shortages and rising prices in Ghana in 1972 formed a critical prelude to the coup that unseated Busia and led to the period of political maneuvers and flux that threatened to overthrow the government of Arap Moi in Kenya in 1980.

It is ironic but true that among those governments most committed to low-cost food are the "radical" governments in Africa. Despite their stress on economic equality, they impose lower prices on the commodity from which the poorest of the poor—the peasant farmers—derive their incomes. A major reason for their behavior is that they are deeply committed to rapid industrialization: moreover, they are deeply committed to higher real wages for urban workers and have deep institutional ties to organized labor.

We can thus understand the demand for low-cost food. Its origins lie in the urban areas. It is supported by governments, both out of political necessity and, on the part of more radical ones, out of ideological preference. It arises because food is a major staple and higher prices for such staples threaten the real value of wages *and* profits.

Partially confirming these contentions is statistical evidence concerning government controls over the retail price of rice. Taking the presence or absence of retail price controls for rice as a dependent variable, I have taken as independent variables the ideological preferences of the various governments, data as to whether or not rice was an urban staple, and measures of the domestic rate of inflation.[6] Employing these variables in a probit analysis, I secured results suggesting that, insofar as rice is a staple of urban consumption, governments are more likely to subject it to retail price control; the greater the rate of domestic inflation, the more likely were governments to attempt to control the price of rice. Moreover, socialist and Marxist governments were more likely to impose price controls than were governments of no discernible ideological stance; capitalist governments were less likely to do so. I obtained similar results for my analysis of government control over the retail price of maize, with one exception: inflation was not significant. Interestingly, a measure of the concentration of urban dwelling was: the greater the proportion of urban dwellers concentrated in the nation's largest city, the more likely the government was to have retail price controls for maize.

There are thus deep-seated political reasons for governments to seek lower food prices. There are also real limitations on their ability to do so. One limitation is political: insofar as farmers themselves are powerful, they are likely to resist the efforts of governments to lower agricultural prices. Only occasionally, however, are farmers powerful. In West Africa, urban/bureaucratic elites have entered rice farming; where they have done so, they have achieved protected commodity prices and subsidized prices for farm inputs (Pearson et al. 1981). In East Africa, similar elites maintain large-scale wheat farms; they too have employed their political influence to prevent adverse pricing policies (Gerrard 1981). But most farms are owned by members of the peasantry, not the elite; they are small-scale, not large-scale; and the farmers are politically weak, not strong. Rarely, then, are farmers powerful, and most often they are taxed.

Political influence on the part of farmers thus occasionally influences the behavior of governments. A more common influence is the limitation of governmental resources. When lower price levels are imposed on farmers, consumers may face shortages. Indeed, food production tends to be highly price elastic; a necessary corollary to low-price policies in Africa is thus the use of public resources to produce or to import food. But most African governments are poor and are short of foreign exchange. Governments therefore lack the resources to make up the shortfalls resulting from their pricing policies, and this places a major limitation on the degree to which they can lower agricultural prices.

Evidence for these arguments is contained in the recent work of Gerrard (1981). Gerrard finds that when the self-sufficiency price for a given crop lies below the world price, then governments tend to set the domestic price at the self-sufficiency price. They also do so when the self-sufficiency price lies above the world price, as in the case of wheat. Gerrard interprets these results as suggesting that African governments regulate prices in order to secure self-sufficiency.

Rather than interpreting these results as suggesting a systematic search by governments for a social objective, I would instead interpret them as the consequence of the interplay between political pressures upon governments and the constraints that they face. Wheat producers are large-scale farmers; often they are members of the political and administrative elite. They therefore possess political power and they employ that power to secure government pledges for self-sufficiency. This political commitment secures them sheltered markets and production subsidies. "Low cost" producers, however, such as peasant producers of maize, lack political power; they therefore face lowered prices. The major limit to the subsidies for the relatively powerful is the cost to government of production subsidies or of food imports, costs that rise as the government deviates from self-sufficiency. Self-sufficiency for wheat producers therefore represents a political triumph for farmers; for maize producers, it represents a political defeat. For both, the limits to the political outcome are the financial costs to the government. This intrepretation covers the facts, as recorded by Gerrard, but rests on a pressure group, rather than welfare maximizing, theory of public policy.

Pressure groups form but one component of the pluralist model of politics.[7] The second major component is competitive elections.

Clearly, were competitive elections contested by rival parties in Africa, agricultural policy could *not* be so strongly biased against rural dwellers. With less than 10 percent of their population in cities, most nations would contain electoral majorities composed of farm families; and electoral incentives would almost inevitably lead politicians to advocate pro-agrarian platforms in their efforts to secure votes and to win power.

Evidence of the significance of electoral incentives is to be found in Zambia. From 1964–1972, the government of Zambia devoted on average over 70 percent of its capital budget to expenditures in the urban areas. In the years prior to national elections, however, the government reallocated its capital program: over 40 percent of the capital budget was then spent in the rural districts. Moreover, it was in the years prior to elections that major rural development programs were announced: the creation of intensive rural development zones, new credit programs, mechanization schemes, or the decentralization of rural administration. The commitment to rural development was thus tied to the electoral cycle. Having periodically to face a rural constituency, the government periodically recommitted itself to the enhancement of their fortunes (see Bates 1978).

There is thus a tension between the two components of the pluralist model. In the African context, the impact of organized interest groups works to the detriment of agrarian interests, whereas competitive elections work to their advantage. It is of course no secret that competitive party systems are rare in Africa. In recent years the frequency of the return to democratic forms of government among the African states has been more than matched by the frequency of the demise of competitive party systems. There is thus little prospect that electoral incentives will counter the biases produced by interest-group politics. Among the primary losers from political disenfranchisement in Africa have been farmers, for it is they who would form electoral majorities.

Governments as Agencies that Seek to Retain Power

Thus far I have analyzed three major approaches to interpreting the behavior of governments in agricultural markets in Africa. Two I have rejected as inadequate. The third, in fact, seems to account for much of what we observe in the area of food policy in Africa. It explains the political pressures for low food prices. Because it accounts for the pressure for low-priced food, it helps to explain why when governments want more food they prefer to secure it by building more projects rather than by offering higher prices. By the same token, it helps to account for the governments' preference for offering production subsidies rather than higher food prices in order to strengthen incentives for food production.

Nonetheless, the pluralist explanation is also incomplete. Its primary virtue is that it helps to account for the essentially draconian pricing policies adopted by African governments. Its primary limitation is that it fails to explain how governments get away with these policies. How, in nations where the majority of the population are farmers and the majority of the resources are held in agriculture, are governments able to succeed in implementing policies that violate the interests of most farmers? In search of answers to this question, a fourth approach is needed: one that looks at agriculture programs as part of a repertoire of devices employed by African governments in their efforts to secure political control over their rural populations and thus to remain in power.

Organizing a Rural Constituency

We have already seen that adopting policies in support of higher prices for agricultural commodities would be politically costly to African governments. It is important to note that this stance would generate few political benefits as well. From a political point of view, conferring higher prices offers few attractions for politicians, for the benefits would be enjoyed by rural opponents and supporters alike. The benefits could not be restricted exclusively to the faithful and withheld from the politically disloyal. Pricing policies therefore cannot be employed by politicians to organize political followings in the countryside.

Project-based policies, however, suffer less from this liability. Officials can exercise discretion in locating projects; they can also exercise discretion in staffing them. This allows them to bestow benefits selectively upon those whose political support they desire. Politicians are therefore more likely to be attracted to project-based policies as a measure of rural development.

The relative political utility of projects explains several otherwise puzzling features of agricultural programs. One is the tendency to construct too many projects, given the budgetary resources available. A reason for this proliferation is that governments often wish to ensure that officials in each administrative district or electoral constituency have access to resources with which to secure political backing (see, for example, Wells 1974; Dadson 1970). Another tendency is to hire too large a staff or a staff that is technically untrained, thus undercutting the viability of the projects. A reason for this is that jobs on projects—and jobs in many of the bureaucracies involved with agricultural programs, for that matter—are political plums, given by those in charge of the programs to their political followers. State farms in Ghana were staffed by the youth brigade of the ruling Convention Peoples' Party, and the cooperative societies in Zambia were formed and operated by the local and constituency-level units of the governing party, to offer but two examples of the link between staffing and political organization. A last tendency is to have projects that are privately profitable but socially wasteful. Again and again, from an economic point of view, agricultural projects fail; they often fail to generate sufficient earnings to cover their costs, and when they do so, they often fail to generate a rate of return comparable to that obtainable through alternative uses of government funds. Nonetheless, public agencies revive and reimplement such projects. A major reason is that public officials are frequently less concerned with using public resources in a way that is economically efficient than they are with political expedience. If a project fails to generate an adequate return on the public investment but nonetheless is privately rewarding for those who build it, provision it, staff it, or hold tenancies in it, then political officials may support it. The project will serve as a source of rewards for their followers and as an instrument for building a rural political organization.

Disorganizing the Rural Opposition

We have seen that government policies are often aimed at establishing low prices for agricultural products. Particularly in the market for cash crops, governments maintain monopsonistic agencies and use their market power to lower product prices, thereby imposing deprivations on all producers. What is interesting, however, is that they return a portion of the resources they exact to selected members of the farm community. Some of the earnings taxed from farmers are returned to a privileged few in the form of subsidies for farm inputs. While imposing collective deprivations, governments thus confer selective benefits. The benefits serve as "side payments": they compensate selected members of the rural sector for the losses they sustain as a consequence of the governments' programs. They thereby make it in the private interests of particular members of the rural sector to abide by policies that are harmful to rural dwellers as a whole. By so doing, they secure the defection of favored farmers from a potential rural opposition and insure their adherence to a governing coalition that implements agricultural programs harmful to farming as a whole.

We have already noted that agricultural producers are both subsidized and taxed. What is of concern at this point is the use of subsidy programs for political purposes. In northern Ghana in the late 1970s, for example, subsidized credit was given to large-scale, mechanized producers who were close allies of the ruling military government. In Senegal, the rural base of the governing party is dominated by the Mourides, a religious sect that earns much of its income from the production of groundnuts; its adherence to the government in power, and to the government's pricing policies, is in large part secured by conferring massive amounts of subsidized credit, land, machinery, and other farm inputs upon its leaders (Cruise O'Brien 1971). In Zambia, access to subsidized inputs could best be obtained by most rural dwellers through membership in agricultural cooperative societies. The societies were formed by local units of the governing party and are now dominated by them; access to inputs is therefore contingent upon political loyalty. The rural loans program, moreover, was run and staffed at the local level by former party "militants" who helped to insure that the "fruits of independence" were given to those who contributed to the cause of the party in power. In Ghana, to cite one last example, the collective resistance of cocoa producers to low cocoa prices in the 1950s was broken in part by the "secret weapon" of the Convention People's Party, the notorious United Ghana Farmers' Council. By distributing pesticides, cutlasses and other "farm inputs" to those who would support the government and its policies, and by restricting access to credit to the political faithful, the Farmer's Council helped to break the resistance of the farming population to the government and its agricultural programs (Ghana, 1956, 1967).

It should be noted, incidentally, that the bestowal of privileged access to farm inputs was a technique also employed by the colonial governments. The exchange of political loyalty for access to these inputs was widely recognized to be part of the bargain. In Northern and Southern Rhodesia, for example (now Zambia and Zimbabwe), the colonial governments used revenues secured by their monopsonistic maize marketing agency to subsidize the costs of inputs that they then lavished upon a relatively small number of "improved" or "progressive" farmers. The nationalist movements presciently labeled these farmers "stooges" of the colonial regimes. They saw that the apportionment of the inputs had been employed to separate the interests of these privileged farmers from the interests of the mass of rural producers and to detach their political loyalties from those of their fellow Africans.

In conferring selective benefits in the markets for farm inputs while imposing collective deprivations in the markets for products, governments secure the deference of a privileged few to programs that are harmful to the interests of most farmers. By politicizing their farm programs and making access to their benefits contingent upon political loyalty, governments secure acquiescence to those in power and compliance with their policies. The political efficacy of these measures is underscored by targeting the large producers who have the most to gain from a change in pricing policy, and who might otherwise provide the "natural leadership" for efforts on the part of farmers to alter the agricultural policies of their governments.

Markets as Instruments of Political Organization

As part of their development programs, African governments intervene in markets to alter prices. At least in the short run, market intervention establishes disequilibrium prices, i.e., prices at which demand exceeds supply. Such prices artificially induce scarcities, and the allocation of these scarce resources through regulated market channels becomes a significant source of political power. Regulated markets can be used to organize political support and to perpetuate governments in power.

As we have seen, in the markets for agricultural products public monopsonies depress the price of commodities below the market price. Those in charge of the market can then bestow the right of entry;

persons given access to the market can reap excess profits and owe their special fortunes to the favor of those in charge. Members of the Cocoa Marketing Board in Ghana, for example, frequently allowed private trading on the part of persons whose political backing they wished to secure. These included members of the very highest levels of the Ghanaian government. In the Kenyan maize market, moreover, the issuance of movement permits by the director of the Maize and Produce Marketing Board was used to create an indebted and loyal political following (Kenya 1966). Granting access to a market in which the price of commodities has been artificially lowered as a matter of government policy thus becomes a valuable instrument in acquiring political influence.

Disequilibrium product prices also facilitate political control by yielding the capacity to *dis*organize those most hurt by the measure: the farmers themselves. For a portion of the gains, the bureaucrat in charge of the market can turn a blind eye while farmers make sales at market-clearing prices. The structure of regulation vests legal powers in the bureaucrats; the farmers have no right to make such sales. Only by securing an individual exception to the general rule can the farmer gain access to the market-clearing price. Within the framework established by marketing policy, the farmers thus do best by securing individual exceptions. The capacity for discretion therefore allows the bureaucrat to separate the individual interests of particular producers from the interests of their class, and collective organization on the part of rural producers becomes more difficult. In addition, the structure of regulations creates for the government essential elements of political power. By allowing exceptions to the rules, the bureaucracy grants favors; by threatening to enforce the rules, it threatens sanctions. Market regulations thus become a source of political control, and this, in a sense, is most true when they are regularly breached.

Governments establish disequilibrium prices in the markets for inputs as well; the result, once again, is the enhancement of their capacity for political control. When they lower the price of inputs, private sources furnish lesser quantities, users demand greater quantities, and the result is excess demand. One consequence is that the inputs acquire new value; the administratively created shortage creates an economic premium for those who acquire them. Another is that, at the mandated price, the market cannot allocate the inputs; they are in short supply. Rather than being allocated through a pricing system, they must be rationed. Those in charge of the regulated market thereby acquire the capacity to exercise discretion and to confer special benefits upon those whose favor they desire.

It is these dynamics that render farm input programs so potent a source of political patronage. On occasion, governments place political "heavyweights" in charge of these programs. The result often is that the elite then consume the rental premium; they sell the inputs at the price they can command in the market. By allowing the corruption of farm programs the governments thus secure the fealty of potent political figures. In other cases, governments forbid corruption and instead allocate the inputs at their officially mandated prices. The result is to secure political loyalty from lower-level political figures—the intended clients and beneficiaries of the subsidy program, for it is they who then secure the economic premium.[8] Moreover, because of excess demand, those who distribute the inputs can make demonstrations of political loyalty a prerequisite for their allocation. Thus do public programs that distribute farm credit, tractor hire services, seeds, and fertilizers, and bestow access to government-managed irrigation schemes and public lands become instruments of political organization in the countryside of Africa.

Conclusions

Governments in Africa, like governments elsewhere in the developing world, intervene in agricultural markets in a way that violates the interests of most farmers. They tend to adopt low price policies for farm products; they tend to increase the prices farmers must pay for the goods they consume. And though they subsidize the prices of goods that farmers use in production, the subsidies are appropriated to the richer few. In addition, the farm policies of African governments are characterized by a stress on projects rather than prices; by a preference for lowering farm costs rather than increasing farm revenues when price policies are used; and by widespread economic inefficiency.

We have examined four approaches to the analysis of policymaking by governments and assessed their ability to account for these characteristic features of African farm policy. Two of the approaches appear to have performed well. Policy formation clearly reflects the desire of African governments to placate organized domestic interests. It also reflects their desire to stay in power. By contrast, two theories did not perform well when evaluated in terms of their ability to account for this collection of "facts." One was derived from radical political economy; the other from neoclassical economics. The first failed to accord adequate political power to the domestic political forces of private interests or governments. The second overstated the importance of the public interest as a determinant of political choice.

Acknowledgments

Research for this essay was supported by the Division of Humanities and Social Sciences of the California Institute of Technology and by the National Science Foundation (Grant No. SOC 77-08573A1). The paper draws extensively on materials presented in Bates (1981, 1983).

Notes

1. In his discussion (see p. 223), Bruce Gardner correctly notes the paucity of references to the literature on industrial regulation. Appearances to the contrary, I am well aware of this literature; one of my goals in this work is to define the problem of agricultural pricing in a way that would lend itself to analysis as a problem of industrial regulation. The most rapid progress in this field will, I feel, most likely be made by those who examine the behavior of marketing boards and other public agencies in Africa from the point of view of those who have analyzed the FTC, the FDA, and other regulatory agencies in the United States.

2. For introductions, see Amin (1973), Baran (1960), Frank (1972), Rodney (1976), and Wallerstein (1974).

3. Some of the most interesting material on this subject is contained in the volumes of the Coker Commission of Enquiry in Western Nigeria. The commission documents the transfer of funds from the marketing to development corporations and banks specifically set up to finance local capitalists seeking to compete with foreign firms (Nigeria 1962).

4. For a critique of dependency theory from the left which makes many similar points, see Warren (1973).

5. In labeling this approach pluralist, I am *not* subscribing to the normative correlatives of that approach, which assumes costless organization and the free entry into the policymaking arena of all potential interest. See Dahl (1961) and Olson (1971).

6. The ideological preferences were assessed by Young (1982). Measures of inflation and urbanization were taken from IBRD (1981). Measures of government market intervention were taken from USDA (1980).

7. A more systematic analysis of pressure group politics, based upon more rigorous microeconomic foundations, is contained in Bates and Rogerson (1980) and Bates (1981). The first deals primarily with the demand for price intervention; the second with the supply. Both attempt to explain the relative inefficacy of farmers in pressure group politics in the developing areas.

8. In Zambia, one of the cooperative societies I studied in Luapula purchased subsidized fertilizers. It then reaped the rental premium associated with the subsidy by selling the fertilizer at market-clearing prices to the local commercial farmers.

References

Amin, Samir. *Accumulation on a World Scale*. New York: Monthly Review Press, 1973.

Baran, Paul. *The Political Economy of Growth*. New York: Prometheus Paperbacks, 1960.

———. *Markets and States in Tropical Africa: The Political Basis of Agricultural Policies*. Berkeley and Los Angeles: University of California Press, 1981.

Bates, Robert H. *Rural Responses to Industrialization: A Study of Village Zambia*. New Haven and London: Yale University Press, 1978.

———. *Essays on the Political Economy of Rural Africa*. Cambridge: Cambridge University Press, 1983.

Bates, Robert H., and William P. Rogerson. "Agriculture in Development: A Coalitional Analysis." *Public Choice* 35, 5 (December 1980):513-527.

Bauer, P. T. *West Africa Trade*. London: Routledge and Kegan Paul, 1964.

Beckman, Björn. *Organizing the Farmers: Cocoa Politics and National Development in Ghana*. New York: Holmes and Meier, 1976.

Cohen, John, and Dov Weintraub. *Land and Peasants in Imperial Ethiopia: The Social Background to Revolution*. Assen: Van Gorcum, 1975.

Cruise O'Brien, Donal B. *The Mourides of Senegal: The Political and Economic Organization of an Islamic Brotherhood*. Oxford: Clarendon Press, 1971.

Dadson, Alfred John. "Socialized Agriculture in Ghana, 1962-1965." Ph.D. Dissertation, Department of Economics, Harvard University, 1970.

Dahl, Robert. *Who Governs?* New Haven and London: Yale University Press, 1961.

Dodge, Doris Jansen. *Agricultural Policy and Performance in Zambia*. Berkeley, Calif.: Institute of International Studies, 1977.

Ekhomu, David Onaburekhale. "National Food Policies and Bureaucracies in Nigeria: Legitimization, Implementation, and Evaluation." Paper presented at the African Studies Association Convention, Baltimore, Maryland, 1978. Mimeographed.

Frank, André Gunder. *Capitalism and Underdevelopment in Latin America*. New York: Monthly Review Press, 1972.

Gerrard, Christopher David. "Economic Development, Government Controlled Markets, and External Trade in Food Grains: The Case of Four Countries in East Africa." Ph.D. Dissertation, University of Minnesota, August 1981.

Ghana. Jibowa Commission. *Report of the Commission of Enquiry into the Affairs of the Cocoa Purchasing Company*. Accra: Government Printer, 1956.

———. *Report of the Commission of Enquiry into the Local Purchasing of Cocoa*. Accra: Government Printer, 1967.

Girdner, Janet, and Victor Olorunsula. "National Food Policies and Organizations in Ghana." Paper presented to the annual meeting of the American Political Science Association, New York, 1978.

Gotsch, Carl, and Gilbert Brown. "Prices, Taxes and Subsidies in Pakistan Agriculture, 1960-1976." Staff Working Paper No. 387. Washington, D.C.: World Bank, 1980.

Griffin, Keith. *The Green Revolution: An Economic Analysis*. Geneva: United Nations Research Institute, 1972.

Helleiner, Gerald K. *Peasant Agriculture, Government, and Economic Growth in Nigeria*. Homewood, Ill.: Richard D. Irwin, 1960.

Hill, Frances. "Experiments with a Public Sector Peasantry." *African Studies Review* 20(1977):25–41.

Huxley, Elspeth. *No Easy Way: A History of the Kenya Farmers' Association and Unga Limited*. Nairobi: Private Printing, 1957.

International Bank for Reconstruction and Development (IBRD). *Kenya: Into the Second Decade*. Washington, D.C., 1975.

————. *Ivory Coast: The Challenge of Success*. Washington, D.C., 1978.

————. *Accelerated Development in Sub-Saharan Africa: An Agenda for Action*. Washington, D.C., 1981.

Jansen, Doris J. "Agriculture Pricing Policy in Sub-Saharan Africa in the 1970s." December 1980. Mimeographed.

Jones, William O. "Agricultural Trade Within Tropical Africa: Historical Background." *Agricultural Development in Africa: Issues of Public Policy*, ed. Robert H. Bates and Michael F. Lofchie. New York: Praeger Publishers, 1980.

Kaneda, Hiromitsu, and Bruce F. Johnston. "Urban Food Expenditure Patterns in Tropical Africa." *Food Research Institute Studies* 2(1961):229–275.

Kenya. *Report of the Maize Commission of Inquiry, June 1966*. Nairobi: Government Printer, 1966.

Killick, Tony. *Development Economics in Action*. New York: St. Martin's Press, 1978.

Kline, C. K., D.A.G. Green, Roy L. Donahue, and B. A. Stout. *Industrialization in an Open Economy: Nigeria 1945–1966*. Cambridge: Cambridge University Press, 1967.

Krishna, Raj. "Agricultural Price Policy and Economic Development." In *Agricultural Development and Economic Growth*, ed. M. Southworth and Bruce F. Johnston. Ithaca, N.Y.: Cornell University Press, 1967.

Leonard, David K. *Reaching the Peasant Farmer: Organization Theory and Practice in Kenya*. Chicago and London: University of Chicago Press, 1977.

Leubuscher, Charlotte. *Bulk Buying from the Colonies: A Study of the Bulk Purchase of Colonial Commodities by the United Kingdom Government*. London: Oxford University Press, 1956.

Lipton, Michael. *Why Poor People Stay Poor: Urban Bias in World Development*. Cambridge, Mass.: Harvard University Press, 1977.

Nigeria, Federal Ministry of Information. *First Report of the Anti-Inflation Task Force*. Lagos: Government Printer, 1978.

————. Federal Ministry of Information. *Second and Final Report of the Wages and Salaries Review Commission, 1970–71*. Lagos: Ministry of Information, 1971.

————. *Public Service Review Commission*. Lagos: Government Printer, 1974.

————. *Report of the Coker Commission of Enquiry into the Affairs of Certain Statutory Corporations in Western Nigeria*. Lagos: Government Printer, 1962.

Njonjo, Apollo I. "The Africanization of the 'White Highlands': A Study in Agrarian Class Struggles in Kenya, 1950–1974." Ph.D. dissertation, Princeton University, Princeton, N.J., 1977.

Olson, Mancur. *The Logic of Collective Action.* Cambridge, Mass: Harvard University Press, 1971.

Onitiri, H.M.A., and Dupe Olatunbosun. *The Marketing Board System.* Ibadan: Nigerian Institute of Social and Economic Research, 1974.

Pearson, Scott R., Gerald C. Nelson, and J. Dirck Stryker. "Incentives and Comparative Advantage in Ghanaian Industry and Agriculture." Paper for the West African Regional Project, Stanford University, Calif., 1976. Mimeographed.

Pearson, Scott R., J. Dirck Stryker, and Charles P. Humphreys. *Rice in West Africa.* Stanford, Calif.: Stanford University Press, 1981.

Pick, Franz. *Pick's Currency Yearbook, 1976–1977.* New York: Pick Publishing Corporation, 1978.

Rodney, Walter. *How Europe Underdeveloped Africa.* Dar es Salaam: Tanzania Publishing House, 1976.

Roider, Werner. *Farm Settlements for Socio-Economic Development: The Western Nigerian Case.* Munich: Weltforum Verlag, 1971.

Schatz, Sayre P. *Nigerian Capitalism.* Berkeley, Los Angeles, and London: University of California Press, 1979.

Schmidt, Guenter. "Maize and Beans in Kenya: The Interaction and Effectiveness of the Informal and Formal Marketing Systems," Occasional Paper No. 31. Nairobi, Kenya: Institute for Development Studies, University of Nairobi, 1979.

Schultz, T. W. *Transforming Traditional Agriculture.* New York: Arno Press, 1976.

Schultz, T. W., ed. *Distortion of Agricultural Incentives.* Bloomington: Indiana University Press, 1978.

Sklar, Richard L. "The Nature of Class Domination in Africa." *Journal of Modern African Studies* 17(1979):531–552.

Stryker, J. Dirck. "Ghana Agriculture." Paper for the West African Regional Project, 1975. Mimeographed.

Swainson, Nicola. *Foreign Corporations and Economic Growth in Kenya, 1977.* Manuscript.

Uganda, Treasury Department. "Statement of Cotton Price Assistance Fund at 31st October 1977." 11 November 1977. Information Sheet.

U.S. Agency for International Development (USAID). *Development Assistance Program FY 1976–FY 1980, Ghana,* Vol. 4, Annex D. (Agricultural Sector). Washington, D.C.: USAID, 1975.

U.S. Department of Agriculture. *Food Problems and Prospects in Sub-Saharan Africa.* Washington, D.C.: USDA, 1980.

U.S. General Accounting Office. *Disincentives to Agricultural Production in Developing Countries.* Washington, D.C.: GAO, 1975.

Van Velzen, H.U.E. "Staff, Kulaks and Peasants." *In Socialism in Africa,* ed. Lionel Cliffe and John Saul, Vol. 2. Dar es Salaam: East African Publishing House, 1973.

Walker, David, and Cyril Ehrlich. "Stabilization and Development Policy in Uganda: An Appraisal." *Kyklos* 12(1959):341–53.

Wallerstein, Emmanuel. "Dependence in an Interdependent World: The Limited Possibilities of Transformation Within the Capitalist World Economy." *African Studies Review* 17(April 1974a):1–26.

————. *The Modern World System.* New York: Academic Press, 1974b.

Warren, Bill. "Imperialism and Capitalist Industrialization." *New Left Review* 81(1973):3–44.

Wells, Jerome C. *Agricultural Policy and Economic Growth in Nigeria, 1962–1968.* Ibadan: Oxford University Press for the Nigerian Institute of Social Science and Economic Research, 1974.

Young, Crawford. *Ideology and Development in Africa.* New Haven and London: Yale University Press, 1982.

6
Why Do Governments Do What They Do? The Case of Food Price Policy

Alain de Janvry
Department of Agricultural and Resource Economics
University of California, Berkeley

The Role of the State in Determining Food Prices

Free-market determination of food and agricultural prices is largely a myth. In essentially all countries, even in those most dedicated to the rule of free enterprise—both less and more developed—these prices are highly influenced by state intervention. Controls can be *direct,* through farm price supports or consumer price ceilings, forced deliveries to governments at mandated prices, export controls and taxes, import tariffs or subsidies, and input price subsidies; or *indirect,* through exchange rate policies.

In a recent survey of 50 developing countries, Saleh (1975) found that in at least 46, government policies undervalued agricultural commodities and created serious disincentives to production. Policy interventions are no less pervasive in the more developed countries, if applied usually in the opposite direction, resulting in a general overvaluation of agricultural commodities (Schultz 1978; Bale and Lutz 1979). Peterson (1979) thus calculated that the terms of trade for agriculture were more than twice as favorable in a sample of 23 more-developed countries than in 30 less-developed countries. The result has been a tendency toward overproduction and underconsumption of agricultural goods in the former countries. Hidden behind these general tendencies is, of course, a tremendous heterogeneity among countries, crops, and time periods.

The consequence of this pervasive involvement of government in price policymaking is that both successes and failures in the performance of agriculture have been attributed to the state. For example, successes with agricultural development in the United States and India have been attributed to the policies of price support and public agricultural research, while the failure of food production to follow effective demand in Egypt and most Latin American and African countries has been blamed on price disincentives resulting from the implementation of cheap food policies. At the level of specific crops, the success of rice production and the failure of wheat in Colombia, the success of wheat and the failure of corn in Mexico, and the success of wheat and the failure of pulses in India have all been attributed to differential price policies among crops.

Since price policy is important in determining the prevailing food and agricultural prices for particular crops in different countries, we need to relocate the study of agricultural prices out of a pure theory of markets into a *theory of the state*. Why do governments implement such markedly different food and agricultural price policies resulting in sharply contrasted patterns of agricultural and rural development?[1] We need to go beyond the neoclassical equilibrium analysis of markets consisting of observing price distortions, calculating the resulting social costs, and merely calling for their removal. To the contrary, these distortions must be understood in terms of their dynamic historical and social reality if any proposal to amend them is to have political relevance: What did governments achieve or try to achieve by the policies implemented? What were the coalitions of forces that led governments to follow the observed courses of action and what were the resulting distributions of gains and losses?

Although they are tremendously complex to understand and highly historically/geographically specific in each instance, a first step toward answering these questions is to identify some regular patterns in the relation between (1) the economic and social structure of particular societies; (2) the forms, functions, and limits of the state; (3) the making and implementation of food and agricultural price policy; and (4) the economic and social consequences of these policies, both within agriculture (in terms of agricultural and rural development) and in the economy at large. It is the identification of these patterns that will serve as a basis for developing a theory of the political economy of food prices. It is consequently toward this ambitious goal that we attempt here to take some modest first steps. To do this, we will seek to answer the following four groups of questions.

1. How do we explain the behavior of the state in a capitalist society, in particular its forms, functions, and limits? Without engaging in a full-fledged review of the competing theories of the state,

what we need is to extract from these theories some key concepts that can serve as classificatory dimensions to identify repeated patterns in government behavior toward agricultural price policy.

2. What is specific about food and agricultural prices in terms of state policymaking? In particular, how do the origins and the destinations of supply locate different social groups relative to the payoffs of price policy? How does the pricing of agricultural and food products enter in the definition and the resolution of particular economic and political crises that call for state intervention?

3. What are some key hypotheses about the relation between social and economic structure; forms, functions, and limits of the state; and price policy that explain the observed highly contrasted patterns of price policy across countries (especially overvaluation versus undervaluation of agricultural commodities and whether or not a price wedge exists between farm and consumer prices) and across crops.

4. Why do specific governments implement specific price policies? Here we will study the cases of the United States, India, Colombia, and Egypt in reference to the selected elements of the theory of the state, the specificity of policymaking regarding food and agricultural prices, and the hypotheses advanced on the relation between socioeconomic structure, state, and policy.

Elements of a Theory of the State

The key questions that a theory of the state should be able to answer relate to the forms, functions, and limits of the state: Why does the state assume different *forms* (liberal democratic, corporatist, dictatorial, etc.)? What are the *functions* that the state attributes to itself or is forced to endorse, i.e., why do governments do what they do? What are the *limits* to state intervention and, hence, what is the effectiveness of the state in performing its functions? These, indeed, would appear to be basic questions to address. Yet, it is evident that the state has been largely understudied in social sciences of all slants, in spite of the growing importance of the state in social life, particularly in economics since the Keynesian revolution and in politics since the events of 1968 in Europe. Nevertheless, these events have stimulated a growing interest in the subject over the last 15 years.

There does not exist at this stage a unified consistent theory of the state, neither in the classics of the state by Weber or Marx nor in subsequent developments in either the pluralist or Marxist traditions. What we have, instead, are different approaches to the state, each with different purposes and in different historical contexts that complement one another in an eventually illuminating fashion. To a large extent, this is precisely because a theory of the state, which is developed in relation to a characterization of social classes and of the emergence of crises, has to be made in reference to a particular historical context, which thus limits the scope of the theory and severely curtails its predictive power. As a result, theories of the state tend to have eventually strong ex post explanatory power but remain weak in ex ante predictive power.

Having introduced these caveats, let us identify several concepts and approaches to the theory of the state that can be useful for our purpose of understanding observed patterns of food and agricultural price policy.

Structural Characterization of a Society

Since the state must be understood in reference to its social and economic context, the first step toward a theory of the state consists in defining the structural dimensions of particular societies. Characterizing a society is a highly complex task; for our purposes here, I will retain the social class structure, the economic structure, and insertion in the international division of labor.

The Social Class Structure. Here the essential element is to characterize what is the class alliance in control of the state. In terms of agricultural and food policy, there are three important elements of class characterization.

The first is landlord power and the importance of rent. The key issue is to locate the social position of landlords in terms of their relationship with the capitalist class. Are they in a prerepeal of the Corn Laws–type situation where the capitalist class is too weak to control the state on its own and must consequently establish an alliance with the landlord class?[2] In this transitory stage, the landlords contribute to the alliance with political support and release on the market of a labor force (e.g., through the enclosure movement), while capitalists insure landlords of the preservation of land rent through favorable terms of trade for agriculture (Rey 1976). As we will show, this is a situation still characteristic of India today. Or, to the contrary, are the landlords in a postrepeal of the Corn Laws–type situation where the capitalist class has acquired enough political power of its own to do away with an alliance with landlords and thus cancel the rent through lower agricultural prices, allowing the rate of profit to rise in the whole economy? As we have argued elsewhere, the land reform programs of the 1960s in Latin America and Egypt had as a fundamental purpose the destruction of the political power of the landed elites and the elimination of the category of rent from the formation of food prices (de Janvry 1981).

The second element to be analyzed is social dualism in food production: capitalist versus peasant producers. There are here two embedded issues. The first is to characterize the relative social and economic importance of peasant and family farmers versus capitalist and agribusiness farmers. In most instances studied, both social categories are present, and the latter, although a numerical minority, controls the bulk of the land, produces most of the marketed surplus, and has a virtual monopoly in political representation of sectoral interests. The second issue, however, is whether or not this structural dualism is reflected in a dualism in cropping patterns. In some countries—the United States and India, for example—structural dualism is not reflected in cropping pattern dualism since the same crops are found across all farm types. In this case, agricultural price policies will affect both family and capitalist farmers. In Egypt and in most Latin American countries, many crops and sometimes the bulk of staple food production are left to peasants while capitalist farms specialize in the production of agroindustrial goods, luxury foods, and agroexports. In this case (staple) food policy affects the peasant sector differentially, while the political power of capitalist farmers can be devoted to the defense of profits in other lines of economic activity.

The third important characterization concerns the social democratic alliance and the urban constituency of the state. It analyzes the social class structure in terms of the position of the working class and peasantry relative to the class alliance in control of the state. There are two issues here. The first is whether workers and peasants are incorporated into the dominant alliance, allowing them to share through wage and terms-of-trade conditions the productivity gains of industry and agriculture. In this case, the concept of parity income becomes an important element of price policy. Or are workers and peasants rejected in open political opposition and subjected to intense surplus extraction under the form of cheap food policies and cheap labor growth? As we will see, this contrasted social alliance has a corresponding basis in the logic of growth according to the alternative economic structures.

The second issue is to characterize the strength of the urban constituency of the state in the making of food price policy. This constituency includes three politically powerful groups that coincide in their demands for cheap food, even if for different purposes. Urban consumers want higher real wages through lower food prices, a particularly important variable when more than half of their take-home pay is spent on food. Employers want to raise the rate of profit through lower monetary wages allowed by cheaper wage foods. Government employees also want cheap food as they are part of the other two groups, being urban consumers in their own rights and employers as managers of public enterprises.

The Economic Structure. Again, there are many relevant characteristic elements of the economic structure (in particular, the nature of inter-sectoral linkages and the labor intensity of technology), but we will consider only two features: social articulation or disarticulation and alternative sources of economic surplus.

Under social articulation, the market for final consumption goods originates principally in the incomes of wage earners and peasants (de Janvry and Sadoulet 1983). As a result, expansion of the domestic market ultimately has to occur via a rise in real wages and peasant incomes tailored to productivity gains. Under social disarticulation, by contrast, the market for final consumption goods originates in upper incomes derived from rents and profits. Market expansion thus requires concentration in the distribution of income, and the dynamics of growth are based on cheap labor. This, in turn, requires cheap food, as the level of monetary wages is principally determined by the level of food prices at low income levels. In terms of forms of the state, social articulation tends to correspond to democratic governments, while disarticulation leads to authoritarian and militaristic regimes. In terms of class alliances in control of the state, social articulation tends to lead to a broad (if, indeed, always conflictive) alliance between capitalists, workers, and peasants; under social disarticulation, workers and peasants have marginal access to the state, and the alliance in power is limited to the landlord and capitalist classes.

The second feature concerns alternative sources of economic surplus. Industrial development requires the mobilization of an investable surplus from other economic sectors. As Kuznets (1964) has argued, agriculture is the sector from which this surplus must originate unless there exists a rich primary export base or other sources of revenue like merchant activities or international transfers of funds. Transfers out of agriculture occur voluntarily through the investment of agricultural savings and rents outside of agriculture and coercively through land, export, and income taxes. But the most common transfer mechanism, particularly for domestically consumed commodities, has been the undervaluation of agricultural output. We thus have a situation where the more forcefully a country wants to industrialize and the less the availability of nonagricultural sources of surplus, the more extractive agricultural price policy will tend to be.

Insertion in the International Division of Labor. Here the two key features are the degree of food dependency and the balance-of-payments position of the country. The first determines, for example, how sensitive domestic prices are to fluctuations in international market prices and in the exchange rate. The second determines the possibilities of making up for deficient domestic production via imports. It also defines the capacity for the state to let a price wedge develop between consumer and producer prices and to make up for the difference out of public revenues.

Two Complementary Approaches to the State

There are fundamentally two contrasted ways in which the state has been conceptualized in the framework of political economy. These two approaches have not been unified in a consistent framework, but they reinforce each other in illuminating the forms, functions, and limits of the state from two different angles.

The Instrumentalist or Class-Political Analysis of the State. This approach looks at the state in terms of class domination (Poulantzas 1975): the state is both an *object* of political conflicts, where the purpose of struggle is to appropriate its institutions and the rents they distribute, and an *instrument* of domination in using control of the state to redefine its forms and functions. It is thus the constellation of political forces in society that determines how the state becomes an instrument to satisfy group interests via political power. In its most operational form, the approach focuses on a class-based analysis of pressure groups that compete for access to and influence over the state bureaucracies. Deprived of its class basis, this analysis of pressure groups reduces to the pluralist logic of collective action (Olson 1965). A key to this analysis is thus to define how different competing interest groups are organized, the extent of their economic and political power, and their ability to gain control of particular aspects of policymaking in defense of their own advantages. In some cases, this results in appropriating the state apparatus in its totality, allowing the making of consistent policies for the group in power; in others it leads to control of specific branches of government by rival interest groups, leading to a veritable balkanization of the state.

An important aspect of this approach is to delineate the realm of individual versus collective initiatives in the quest for economic advantages. Individuals are, indeed, always faced with the alternative of either trying to outcompete or outwit other members of the same class, for example in deciding on resource allocation or technological alternatives at the level of the firm, or of organizing collective action, for example to modify the price conditions faced in markets or the rules for access to additional resources. In general, collective action tends to be in response not so much to incentives but to coercion occurring in the context of economic or political crises. In that sense, crises are important focusing devices that stir class mobilization and guide the direction of collective action (Hirschman 1958).

The Capital Logic Analysis of the State. A major difficulty with the instrumentalist analysis of the state is that it fails to ground state action in the systemic logic of growth, both in terms of the requirements of the process of capital accumulation and the preservation of the social relations of capitalism. As a result, the bounds within which social conflicts occur over and through the state are not clearly delineated: the approach is mainly concerned with distributional struggles, and the boundaries of these struggles are left unspecified.

The capital logic approach, by contrast, explains the need for and the behavior of the state in terms of the economic and political needs of capital to insure both sustained economic growth and reproduction of the social relations of capital (Baran and Sweezy 1966). The key idea is that there are collective interests to capital that cannot be insured by individual capitalists alone. This creates the need for the state to act as an ideal collective capitalist. And since this need becomes recognized in the context of economic and political crises, it is in those circumstances that the forms, functions, and limits of the state are redefined.

The functions of the state are established in relation to the needs of capital accumulation and of the reproduction of the social relations of capital. These collective functions of the state imply that it is relatively *autonomous* from specific segments of the dominant class alliance. At the same time, the state is both *excluded* from the direct generation of surplus that forms the basis of a fiscal budget and *dependent* upon surplus generation in civil society if its budget, and the salaries of politicians and civil servants, is to be maintained. As a result, successful management of the process of accumulation by the state itself is a precondition for continued existence of the state and of the vested interests it contains. Indeed, the most binding limit of the state is the occurrence of fiscal crises that are themselves determined by accumulation crises (O'Connor 1973). The other limits of the state are found in the legitimacy of the state apparatus and in its administrative capabilities to handle specific forms of crises.

The main contribution of the capital logic approach to the state is to show that the state is not a mere political instrument of class conflicts. Indeed, this is what the pressure group analysis of the state fails to take into account—that the interests of specific groups, endorsed by the state, are limited by the economic and political requirements of the reproduction of capital as a whole. Yet, the capital logic approach has several shortcomings. One is that it imbues the state with the semblance of excessive logic and power over civil society and fails to account for the internal divisiveness, imperfect information, and limited capacity of the state apparatus. In particular, it fails to acknowledge that the state itself can be a source of crises, which often originate in the contradiction between forms of the state (liberal democracy for legitimation purposes) and functions of the state (keeping wages, especially the social wage, in check to defend the rate of profit and accumulation). The other is that it excessively privileges the collective good over the interests of individual capitalists precisely when the instrumentalist approach was stressing exclusively the redistributive gains of interest groups over the collective interests of the dominant class alliance. Clearly, a balanced approach between these two is necessary to approximate a theory of the state. This is what we propose to do in analyzing the making of agricultural and food price policy.

Specificities of the State in Regimes of Transition

The elements of a theory of the state we outlined above were defined in the context of advanced industrialized countries. Their application to less-developed countries characterized by regimes of transition toward eventually mature capitalist systems requires specification of the features of transitory regimes that affect both the instrumentalist and capital logic performances of the state.

We will retain here only one structural feature of transitory regimes, whatever the type of society they emerge from (but usually either feudal, populist, or state capitalist)—the weakness of the emerging urban industrialists to politically confront the power of their peasantries and working classes. This feature can be used as a definition of transitory regimes, and it is all the more evident in the Third World countries of today where economic growth tends to produce serious disruptions of traditional society (due, for example, to inappropriate technology, social disarticulation, and international dependency) and sharply worsening patterns in the distribution of income. We find, as a result, the emerging capitalist class in need of relying on either one of two strategies:

1. Maintain a political alliance with the landlord or rich farmer class in an exchange of agrarian political support against perpetuation of a land rent paid under the form of favorable terms of trade for agriculture (Rey 1976; Mitra 1977). We will refer to this situation as one of "prerepeal of the Corn Laws" when the state surrenders high agricultural prices to the landlord and rich farmers' lobby at the cost of reduced profit rates in the rest of the economic system, slower economic growth, and reduced real wage earnings.

2. Buy legitimacy from the politically demanding segments of the working class (usually the urban constituency) through extensive welfare assistance, particularly in the form of massive food subsidies. We will refer to this situation as one of "prerepeal of the Poor Laws" where the state incurs high social expenses for the sake of political legitimation. The legitimacy crisis of the capitalist class is thus displaced at the level of the state under the form of high public expenditures. The possibility

for the state to absorb this crisis very much depends on its capacity to mobilize a financial surplus, either from dominated segments of domestic society or from abroad.

Crisis Response Versus Planning Mode

The final element we will retain to analyze food price policy is the role of crises in stirring state action and, hence, in defining the time horizon the state confronts in the definition of policy. If the state fundamentally acts in crisis response, the time it has to solve the crisis at hand is typically short and often further shortened by a lack of correspondence between political time and the economic time needed to develop solutions. In a crisis response mode, most instruments used tend to be redistributive, with price policy a particularly evident instance of a short-run zero sum game (Thurow 1980). If the state acts in a planning mode, defining policy in terms of either longer-run structural changes or the anticipation of crises, then the instruments used can be not only redistributive but also growth promoting. Thus, price policy can become an important instrument to stimulate investment and technological change through the income effects it creates. With the failure of "developmentalist" planning in Third World countries (Healy 1972; Hirschman 1981), especially in Latin America, and the increasing shift toward neoliberal and bureaucratic-authoritarian regimes (Collier 1979), the time horizon of economic policymaking has typically been shortened, planned structural change and crisis anticipation have given way to short-run stabilization policies, and planning modes have given way to crisis response. In the making of price policy, this implies increasing dominance of redistributive over growth purposes and, hence, of instrumentalist over capital logic motives.

Specificity of Agricultural and Food Price Policies

Given these general features of the state, there are also specific aspects which differentiate the issue of agricultural and food price policies from other forms of state intervention. In an instrumentalist sense, one is the particular social constellation that is affected by these price policies. In a capital logic sense, price policy is distinguished by the kinds of economic and political crises to which food prices potentially contribute.

Distribution of Payoffs from Food Price Policy

In the short run, without output response, the payoffs from price policy are determined by whether price changes create a net income or a net cost effect for specific social groups. Income effects are determined by the mapping of cropping patterns over the social class structure, i.e., who produces which crops and for what purpose (home consumption versus sale). In the longer run, when output effects are induced by price changes, income effects for other social groups also result from changes in employment opportunities (Mellor 1978). Cost effects are created by the impact of changing prices on real income and real wage in terms of consumption capacity. In addition, monetary wages are affected in the longer run by changes in food prices, resulting in changes in the rate of profit for employers in the economy at large. The important distinction here is between types of commodities in terms of social and geographical absorption: subsistence goods produced and consumed by peasants, wage foods consumed out of urban and rural wage incomes, luxuries consumed out of upper income levels derived from profits and rents, and exportables. As we saw before, the mapping of production of these commodities over rural social classes indicates whether there exists a dualism in production, whereby peasants/family farmers produce wage foods while capitalist farms produce luxuries and exportables, or whether cropping patterns are relatively homogenous across farm types. In an instrumentalist sense, since the social class structure determines access to and control over the state, this dualism in production is key in explaining differential policies toward types of commodities defined in terms of the social and geographical absorption.

Crises Involving Food Prices

Food prices contribute to the definition and potential resolution of both accumulation and legitimacy—the two main purposes of the state in a capital logic sense. In accumulation crises, the role of food prices is essential in the following six fashions:

1. Because food items are essential wage goods, food prices influence the level of monetary wages and hence the rate of profit.
2. Because food items are important elements in the consumer price index, food prices affect the prevailing rate of inflation.
3. Because the sale of agricultural commodities is the source of agricultural incomes, agricultural prices determine both the level of investment and growth in agriculture and the level of effective demand for other sectors of economic activity originating in the agricultural sector.
4. When agricultural products are exportable and importable commodities, agricultural and food prices contribute to the country's balance-of-payments position.
5. Because agriculture is a source of surplus captured by the state, principally under the form of procurement and export taxes, agricultural and food prices contribute to public revenues.

6. If a price wedge develops with world below domestic or with consumer below producer prices, the state has to make up for the difference through public subsidies.

Agricultural and food prices are also important in the definition and management of legitimacy crises. The prices of wage foods define the levels of real income and wage and thus influence the consumption and nutritional levels of the population. In addition, the farm price of agricultural commodities affects the level and distribution of income in agriculture, and farm prices also affect the level of employment and hence the welfare of the rural landless and marginal farmers. As the prices of the essential elements of basic needs and of the ethical "right" of populations to these needs, food prices strongly determine the legitimacy of the state in power.

The key issue in the management of crises through agricultural and food price policy is the contradiction between legitimacy and accumulation crises. The former calls for cheap food policies and public subsidies, while the second, if it occurs in agriculture, requires the defense of agricultural incomes and the stimulation of investment and technological change. The management of this contradiction is at the essence of a theory of the political economy of food and agricultural price policies. It relates to the origin and relative urgency of legitimacy and accumulation crises, to the performance of the state in a crisis versus a planning mode, and to the availability of resources external to agriculture for the state to manage both legitimacy and accumulation crises in the rest of the economic system. It also relates to the degree of autonomy of the state from narrow class interests and to its ability to perform in a capital logic sense versus merely an instrumentalist fashion.

Hypotheses About the Relation Among Structure, State, and Agricultural and Food Price Policy

We are now equipped with the following elements to advance hypotheses on the contrasted nature of observed food and agricultural price policies in different countries and time periods and for different crops and products. They are:

1. A structural characterization of societies in terms of: social class structure—class alliances and control of the state; economic structure—social articulation or disarticulation and sources of economic surplus; and insertion in the international division of labor—food dependency and balance-of-payments positions.

2. Two complementary approaches to the state: a class political-instrumentalist approach, with an identified social distribution of payoffs from agricultural and food price policies, and a capital logic approach, with a specified role for the management of accumulation and legitimacy crises through food price policies.

3. A specification of states of transition to capitalism in terms of class alliances with either the landlords or the urban constituency.

4. Identification of the contradictions inherent to price policies in the context of legitimacy versus accumulation crises, of instrumentalist versus capital logic demands on the state, and of crisis response versus planning mode in policymaking.

These elements of a model of agricultural and food price policy formation are to be used to explain general overvaluation versus undervaluation of agricultural commodities in particular countries and favorable versus unfavorable price policies toward specific crops within a country. They should also explain the occurrence of public budget deficits in the management of agriculture and food policies due to a price wedge between consumer and farm or import prices. The countries we have taken as case studies—the United States, India, Colombia, and Egypt—illustrate the relevance of many of the key categories we are working with.

The contribution of each of the structural variables selected above on the direction of price policy for staple food items is as follows:

1. Social articulation creates the capital logic for a broad alliance to defend farm prices in relation to productivity growth, for the sake of creating effective demand for the key growth sectors. Social disarticulation implies the capital logic of cheap labor and cheap food.

2. Availability of nonagricultural sources of economic surplus creates the possibility for favorable price policies toward agriculture and for the development of a price wedge between producers and consumers. The unavailability of such sources, which characterizes most resource-poor, less-developed countries, implies that programs of accelerated industrialization occur on the basis of undervalued agricultural commodities through both cheap food policies and export taxes.

3. Unimodal production patterns establish the instrumentalist defense by rich farmers' lobbies of a favorable price policy toward food items. By contrast, a bimodal production structure (with peasants and family farmers producing staple food while rich farmers principally produce other commodities) leads to incipient political defense of the pricing of staple foods. This dualism (a) is generally itself created by cheap food policies but leads to reinforcement of the political possibility of cheap food; (b) is not static and exclusive as peasants can sometimes partially shift their cropping patterns toward the more profitable activities; and

(c) is reinforced by institutional rents whereby subsidized credit, access to technology, and infrastructure are distributed by nonmarket mechanisms.

4. In regimes of transition where the dominant class alliance rests heavily on the landlords and rich farmers for political control, the making of agricultural price policy results in the transfer of rents to agriculture via favorable terms of trade.

5. Regimes of transition that face serious legitimacy crises with their urban constituencies implement cheap food policies at the urban level. If public revenues or foreign aid transfers allow for it, a price wedge may develop between producer and consumer prices.

6. The social and geographical origin of specific agricultural products and the social and geographical destination of that supply create a correspondence between unequal power positions relative to the state (instrumentalist logic), unequal significance of different products relative to accumulation and legitimation crises (capital logic), and unequally favorable or unfavorable price policies for specific products.

Case Studies of the Political Economy of Agricultural and Food Price Policy

The United States

In the United States, the joint structural features of social articulation and unimodal production patterns have meant the rare coincidence of a capital logic possibility for agricultural incomes to increase in relation to productivity growth, and a strong instrumentalist defense of the parity of agricultural incomes through the congressional power of the large farmers' lobbies. As a result, agricultural price policy since the 1930s has been effectively legitimized in the name of concerns with rural development issues such as defense of the family farm, rural communities, and the fundamentalism of agrarian values. The consequences have been a favorable price policy toward agriculture, the unleashing of a dynamic technological treadmill sequence of falling relative prices *following* the cheapening of production costs, and a strong record of land and labor productivity growth. Yet, the rhetoric of price policies in the name of rural development has negated its promises. The family farm has been rapidly undermined, land and production have been concentrated on large farms under the strong domination of agribusiness, the mass of rural population has been expelled to the cities, and farm poverty has been minimized only because of the strong labor absorption capacity of the rest of the economic system and the decentralization of industry allowing important off-farm income opportunities.

Social articulation of the U.S. economy is evidenced by the fact that an important share of final consumption for the key sectors of economic growth originates in the incomes of workers and employees. For instance, the 60 percent poorest (which corresponds to Singleman and Wright's 1978 estimate of the share of households headed by workers and semiautonomous employees in the U.S. population) consume 42 percent of the production of electrical appliances and 35 percent of the production of cars. By contrast, these figures for Mexico are, respectively, 22 percent and 1 percent; for Brazil, they are 5 percent and 2 percent (de Janvry and Sadoulet 1983). It is this feature of social articulation that has made it possible for wages and farmers' incomes to increase in relation to productivity growth. As a result, the purchasing power of the farm sector has, historically, played an important role in expanding the domestic market for U.S. industry. This, in turn, has been allowed for by a policy of cheap food *resulting* from promotion of a developmentalist sequence of favorable terms of trade inducing a technological treadmill in contrast to an extractivist policy that cheapens food but stifles agricultural incomes, productivity growth in agriculture, and expansion of a domestic market for industry out of farm incomes.

Unimodality of production patterns exists even though a wide spectrum of farm sizes is found in the production of all main commodities, although evidently not to the same degree. The result of this unevenness in the presence of large farms across commodities and of the different political power of large farmers across regions is that not all commodities have gained equally favorable terms of trade protection. This is important in the United States because agricultural policy is made not as a comprehensive policy package but fundamentally on a commodity-by-commodity basis in response to the instrumental demands of farmers' lobbies. The southern cotton lobby has, for instance, strongly controlled price policy for that commodity over the years. Yet, we find that strong farmer lobbies have been present for the defense of prices of all major commodities, leading Gardner (1981, p. 74) to conclude that "It is difficult to say which commodities are treated best."

Agricultural price policy in the United States has been the cornerstone of farm policy in the sense that the main purpose of policy has been the instrumental distribution of *income* among interest groups (made possible in a broader capital logic sense by the feature of social articulation). The main instrument used for the defense of income has been to support *prices* on a commodity-by-commodity basis. This has been obtained through the stimulation of domestic (food stamp program) and foreign demand (P.L. 480), the support of farm prices through "loan rates" for cereals at which the state guarantees purchase, direct deficiency payments to farmers to make up the difference between target and market prices, and acreage controls to restrict supply while compensating farmers for the corresponding loss in income. These programs were usually legitimized in the name of protecting family farmers' incomes, even though they were evidently sponsored by large farmers who captured the bulk of the resulting benefits and used them to consolidate their economic and political positions. Thus, Hadwiger and Talbot (1979, p. 23) observed that "the great concentrations of landholdings in the South supported a structure of elites that provided national leaders who helped secure federal policies amenable to large-farm agriculture notwithstanding the national-legal symbolisms favoring small-farm holdings." Instrumentally, this was made possible by the control southern and midwestern interests had been able to establish over key congressional committees legislating farm policy as well as by the co-option of the urban vote under the legitimation of defense of democratic ideals in agriculture (Vogeler 1981).

The combination of favorable terms of trade policies and a developmentalist policy package, including agricultural research and extension as well as infrastructure development and subsidized credit, led to strong achievements in land and labor productivity growth. The very instrumentalist basis of policymaking resulted, however, in a regressive redistribution of income toward large farmers and accelerated the demise of the family farm and the rural community, negating the rhetorical purpose of farm policy without significant political challenge, at least until the early 1970s. Thus, in 1969, the largest 7 percent of farms in terms of volume of sales captured 40 percent of the benefits from commodity programs, while the smallest 51 percent received only 9 percent of the subsidy benefits (Schultze 1971). In addition, the larger farms' share of the subsidies was higher than their share of presubsidy income, with the result that farm programs increased income inequality. These farm programs have also been costly for consumers and taxpayers. Gardner (1981) estimated that, in 1978–1979, 80 percent of the income transfer to the farm sector was paid by consumers through higher prices, while 20 percent was paid by taxpayers as the cost of public programs (p. 73).

The conclusion is that, when rich farmers dominate farm policy in an articulated-unimodal context, the benefits of policy are necessarily captured by a minority of them, and the majority ends up having to move into town. The rhetoric of rural development serves effectively the purpose of agricultural development; the social cost is the accelerated demise of family farms and rural communities.

India

The punch line of agricultural and food price policy in India since 1965 is that it has been successful in stimulating technological change and the production of foodgrains to the point of eliminating reliance on imports, but has not increased per capita consumption nor reduced the percentage of the population below required nutritional intake. This has been the consequence of an agricultural price policy relatively favorable to commercial farmers while consumer subsidies have generally not benefited the most needy majority of the rural population. At the level of the state, this has been the consequence of a strong instrumental domination of the commercial farmers' lobby over the making of agricultural policy and their sharing of government control in an alliance with the technocracy and the owners of monopolistic industrial houses.

In terms of the structural variables we selected for analysis, India fits into the category of disarticulated industrialization. Massive investments in capital goods and intermediate goods made between 1955 and 1975 sustained a respectable average annual growth rate in industrial production of 5.8 percent. With the demand for wage goods limited by massive poverty, industrialization for final consumption has been oriented toward the production of luxury goods. Yet, the production of luxuries remains confined to a narrow market and constrained by monopolistic practices and government licensing restrictions on new investments. The result is that the overall growth rate of national income has, for the last 30 years, "been stagnating around a miserable mean of about 3.5 percent" (Krishna 1981).

The production structure is largely unimodal across farm types, not in the sense that all farm sizes have the same cropping patterns, but at least in that the major food grains (rice, wheat, jowar) tend to dominate land use across all farm sizes. The result is that agricultural price policy toward foodgrains affects commercial as well as peasant farms. Among foodgrains, however, wheat tends to be grown on large commercial farms while rice has traditionally been a peasant crop. Historically, this has meant much more favorable price, credit, technology, and infrastructure policies toward the former, resulting in successful production and productivity growth in wheat and only mediocre gains in rice. In recent years, however, the northwestern commercial farms have increasingly diversified production with wheat as a winter crop and rice as a summer crop, bringing the instrumental basis of wheat and rice price policies in closer correspondence.

Finally, the dominant political coalition in India is, since the Green Revolution in the late 1960s, based on an alliance between landlords and commercial farmers, the monopolistic industrial establishment, and the technocracy (Mitra 1977; Bardhan 1978). In this alliance, the landlords and commercial farmers receive benefits in the form of protected agricultural prices, subsidized inputs, institutional rents (credit, fiscal rebates, infrastructure investments), and postponed land reforms. The industrialists "obtain the prerogative of exercising unfettered jurisdiction over industrial, trade, and licensing policies as well as over the management of foreign exchange and of the monetary and fiscal instrument" (Mitra 1977, p. 170). And the technocrats have privileged access to subsidized food in "fair price shops." In the context of democratic policies, the extremely small industrial class needs to maintain this alliance with landed interests to be part of dominant alliance. The cost to them has been the agrarian rent paid through favorable terms of trade policies toward agriculture.

The agricultural and food policies of India since 1965 have been characterized by a system of farm price support on the one hand and of limited consumer subsidies on the other hand. Price support has been insured through definition of a "procurement price" at which the government purchases all grains offered. This price is fixed annually by the Agricultural Prices Commission (APC) and later marginally adjusted at the level of each state. It is supposed to be determined in relation to both the prevailing average cost of cultivation and some principle of income parity. In a recent study of Indian price policy, Krishna and Chaudhuri (1980) have shown that, since the mid-1960s, the procurement price for wheat has been well above the cost of production and, in some years, very close to the free-market price, itself inflated by the scheme of consumer subsidies. For rice, which has not had the defense of an equally strong farm lobby, the procurement price has been below the cost of production in the southern states and West Bengal, but covered the costs of production in the politically powerful northern states. For India as a whole during the 1970s, the procurement price of rice was, however, generally higher than the weighted average cost of production (Subbarao 1982).

Consumers are protected to some extent through distribution of food at subsidized prices (issue price) in the context of fair price shops to which access is restricted by ration cards. But these shops are only available to a minority of urban consumers who may constitute the most politically demanding segment of the population but not necessarily the one in greatest nutritional need.

The making of agricultural price policy evolved from a capital logic purpose between 1950 and 1965 to an increasingly instrumentalist response thereafter. During Nehru's regime, the crisis provoked by an acute food shortage led to the definition of policies directed at increasing the marketed surplus of foodgrains through forced deliveries (which were not successful) and the import of subsidized P.L. 480 grains (which could not be sustained). At that time, the state was operating in a planning mode with considerable autonomy, and several leaders had visions of national purpose. In the subsequent period, however, after the successful achievements of the Green Revolution, the class alliance in power increasingly atrophied, and the state fell prey to parochial instrumentalist demands enforced on an authoritarian basis. In relation to agricultural price policy, this took the form of lobbies attempting to influence the estimation of costs of production and government procurement decisions. In general, it can be said that the APC has increasingly been dominated by rich farmers' interests and that it has failed to include representation of consumers' interests.

The consequences of policy have been favorable prices, particularly for the commodities dominated by rich farmers. Thus, wheat price policy has allowed high rents, and the fall in wheat prices in recent years has lagged far behind the significant productivity gains from the Green Revolution. The consequences of high farm prices have been substantial income gains for the large farmers who not only produce the bulk of the marketed surplus (18 percent of farm households produce 67 percent of the marketed surplus—Patnaik 1975) but have also been the prime adopters of cost-reducing modern technology. The impact on consumers has been the opposite. Between 1962 and 1974, the consumer price of cereals increased faster than the general level of prices. Incomes of the poor, by contrast, remained essentially stagnant. The result is that per capita consumption of cereals has declined by 14 percent in rural and 9 percent in urban areas. As George (1979) has shown, these observed changes in income and price levels explain more than half of the observed changes in per capita consumption. Success in production with falling per capita consumption has led to the accumulation of substantial stocks of foodgrains and to increasing exports of foodgrains as a palliative to deficient domestic effective demand.

Control of the rich farmers' lobby over agricultural price policy has led to implementation of developmentalist policies. The technological treadmill has allowed prices to fall after costs and thus stimulated the diffusion of modern technology. And the management of policy has been conducted in the planning mode inherited from the Soviet inclinations of Nehru's regime. Particularly in recent years, with increasing visibility of the contradiction between successful growth of production and stagnating effective demand, as well as growing inequalities within agriculture itself, farm price policies have been justified in the name of farm income parity. As a result, there has been an increasing shift away from broad-based land reform and commodity development programs toward narrowly defined agricultural development programs aimed at a small but variable spectrum of commercial farmers.

Colombia

Among major Latin American countries, Colombia is probably the one where the state has remained most exclusively under control of an upper class elite, with strong roots in agriculture, that has shed its ancient feudal privileges and engaged aggressively and often violently in a process of capitalist modernization. The result has been an agricultural policy strongly determined in terms of the direct interests of the landed elites. It has induced economic growth in the lines of production controlled by the large commercial farms. At the same time, exclusion of peasants from the political process, failure of the land reform to redistribute the land, the most unequal distribution of income in Latin America, and an often repressive form of government dominated by the executive and ruling by decree have created a deep social dualism. The peasant sector has been repressed both economically and politically, and the crisis of peasant agriculture has pushed upward the price of staple foods and created a domestic food crisis.

Structurally, Colombia is characterized by social disarticulation, evident in the strongly polarized distribution of income and in the rapid rate of industrial growth dominated by the production of consumption goods for the upper-income levels. In agriculture, there is a strong dualism in production, with a minority of large-scale estates devoted mainly to the production of agroindustrial inputs, luxury foods, and agroexports, while a multitude of peasants produce staple foods on minute plots of land. The result is a sharply unequal political representation of commodity interests. In the struggles over policymaking, price policy has, however, been a relatively secondary instrument compared to the issues of land reform, credit, infrastructure investments, trade, and rural development programs. In contrast to India, institutional rents instead of agricultural prices have been the instrument of income transfers in favor of the landed elite. This is because Colombia is a much more urbanized and industrialized country where the economic interests of the dominant groups are highly diversified out of land assets. Economic gains for them result from the promotion of cheap food policies, either through developmentalist sequences (rice) or, more generally, through extractivist policies toward staple foods. Price support policies, consequently, have been generally ineffective, as prices were set too low and were subsequently exceeded by market prices (Sanders 1980). The level of agricultural prices has been more influenced by global economic policies such as exchange rate and import policies than by direct price manipulations. Key among these has been a tendency for overvaluation of the exchange rate due to favorable balance-of-payment positions, resulting in the cheapening of domestic food, and massive imports of subsidized U.S. wheat in the 1960s, which collapsed domestic prices and deeply undermined domestic production.

The making of agricultural policy occurs through a bargaining process between the producer associations—the "gremios" of coffee growers, sugar producers, rice farmers, cattlemen, etc.—and the technocracy and the executive. In the bargain, the first push their instrumental demands for commodity-specific economic advantages, while the second presumably reconcile those in the capital-logic sense of dealing with the requirements of both accumulation and legitimation. In the balance of power, however, it is clear that at least some of the gremios have sufficient weight to undermine the formulation of a consistent policy package. The legislature, for all practical purposes, has been removed from the policy process. Peasants have simply never gained access to decision-making and are reduced to expressing their demands outside institutional channels through, for example, land invasions and strikes. Since the 1950s, by contrast, international lending agencies have had considerable influence over the making of domestic policies—often in the direction of promoting the capital-logic legitimation functions that national governments, hard pressed with instrumentalist demands, have failed to assume. This has been particularly evident with the financing of the schemes of integrated rural development (DRI) and nutritional assistance (PAN) in the 1970s. Due to the strong dualism in production, rural development programs have been managed separately from the mainstream of agricultural policy. They have acquired a fundamentally legitimating purpose as they were organized as a substitute for land reform once the land reform program had effectively fulfilled its antifeudal objective but had, evidently, failed to eliminate rural poverty.

Agricultural policy has been strongly developmentalist for those commodities underwritten by strong associations of large producers, particularly coffee, sugar, rice, cattle, and cotton. The case of rice is remarkable since the combination of protective trade policies, government subsidies, and technological change has unleashed one of the most successful technological treadmills in Latin America (Scobie and Posada 1978). As a result, rice is one of the few staple foods produced profitably in commercial agriculture. At the same time, policy has been extractivist or negligent of peasant agriculture, destroying it through land reform laws that repeatedly forbade sharecropping and tenancy arrangements and marginalizing it from access to supportive institutions (Kalmanovitz 1978). As we said before, this has resulted in the strong growth of the crops produced by large-scale agriculture (cotton, sugar, rice, soybeans, and sorghum) and the stagnation of those produced by peasants (raw sugar, beans, corn, wheat, and potatoes). With agricultural policy dominated by the lobbies of an agrarian elite, the outcome has been successful agricultural development in the production of agroexports, agroindustrial inputs, and rice. It has failed to insure rural development—with the consequent engineering of a domestic food crisis and strong inflation in staple food prices—and has failed to substantially improve the nutritional standards of the mass of the population.

Egypt

Agricultural and food price policy in Egypt today is the contradictory product of the populist (Arab socialism) policies introduced by Nasser and of the policies of economic opening (*infitah*) initiated under Sadat. It combines strong government control over and surplus extraction from peasant agriculture, massive food subsidies to the benefit of consumers, and the emergence of a new class of capitalist farmers who largely escape government controls. The upshot has been stagnant production of traditional crops, rapid increase in food dependency, and expenses on food subsidies exhausting 12 to 15 percent of the public budget.

In terms of the structural categories used in this paper, Egypt's growth pattern is characterized by social disarticulation. Growth has, in recent years, been rapid but directed at the production of capital goods (Soviet planning), luxury goods (including upper-income urban construction), and services. This growth has been stimulated by a favorable foreign exchange situation (due particularly to oil exports, U.S. foreign aid, and remittances from Egyptian workers abroad) but has remained of the wrong type in terms of both creation of self-sustaining domestic employment and satisfaction of basic needs beyond food. As in the model of disarticulated growth, it remains based on cheap labor and cheap food.

Agriculture itself is bimodal in terms of production patterns. Although the land reform of 1952 eliminated feudal estates, it only redistributed some 13 percent of the land and resulted in a new minority class of capitalist farmers with strong political power (Radwan and Lee 1979). The mass of peasant producers are strictly controlled through forced affiliation to supervised cooperatives. These cooperatives control the delivery of inputs (water, fertilizers, credit); the pattern of land use (land quotas in certain crops); and the mobilization of agricultural surpluses (forced deliveries at prices below both world and domestic residual free-market levels). In spite of input subsidies, the result has been a substantial net resource transfer out of agriculture estimated at 5 to 22 percent of agricultural income. Rich farmers, however, escape these controls by producing uncontrolled items such as fruits, vegetables, soybeans, berseem, and meat, and they appropriate the lion's share of credit and subsidized inputs. The result on the distribution of income in agriculture has been regressive.

Finally, although Egyptian agriculture is either negatively protected (traditional food crops and cotton) or in a free-trade regime (new crops), the distribution of staple foods (wheat, rice, cooking oil) is heavily subsidized, particularly in the urban sector. These subsidies derive from a long heritage of Islamic communitarian solidarity and from populist redistributive measures initiated by Nasser that subsequent regimes were unable to remove. They have produced the remarkable result of virtually eliminating malnutrition in Egypt in spite of abysmally low income levels for a majority of the population.

Farm prices for the traditional food crops and cotton, as well as consumer prices, are determined by a constellation of ministries with relatively weak coordination. The result is a system of prices that is far from internally consistent, whose contradictions are rejected at the level of the state and absorbed under the form of rapidly growing public budgets. Initially, the dominant objective of farm price policy was the capital-logic purpose of mobilizing a surplus out of agriculture (especially through cotton and rice exports and domestic cheap food), to sustain rapid heavy industrialization (the Soviet model). However, agriculture has, in the 1970s, been replaced by oil, foreign aid, and worker's remittances as important sources of investable funds and foreign currency. In the process, agriculture has lost its leading sector role. Yet, surplus extraction out of peasant agriculture remains unabated due to lack of political power for that class, although capitalist agriculture progresses in the production of (instrumentally well-defended) new crops. The making of agricultural and food price policy in Egypt should thus be understood in terms of (1) the capital-logic purpose of facing a strong legitimacy crisis at the level of the urban constituency (the food riots of 1976) through surplus extraction out of peasant agriculture, extensive food subsidies, and massive public budget costs and (2) the instrumental alliance among bureaucracy (including the military); the new *infitah* class in the industrial, merchant, and services sectors; and the rich capitalist farmers resulting in the latter capturing favorable terms of trade for their new crops and generous institutional rents.

One consequence of this price policy has been a strong undervaluation of staple foods and cotton. For instance, the effective rate of protection between 1965 and 1970 has been, on the average, 0.76 for rice and 0.58 for cotton (Cuddihy 1980). As stagnation in the yields of cotton and rice clearly indicates, cheap food has been obtained through extractivist policies. Prices have been forced down ahead of costs, resulting in stagnation of domestic production and increasing food dependency. Thus, in 1980, the self-sufficiency ratio of Egypt was only 24 percent for wheat, 26 percent for edible oils, 57 percent for sugar, and 10 percent for lentils, with an overall self-sufficiency ratio of 72 percent (Alderman 1982). Clearly, Egyptian price policy has increasingly been implemented in a crisis mode, responding to a legitimacy crisis through cheap food policies that could only condemn long-run Egyptian food security. In the short run, the crisis can be displaced at the level of the state through massive imports and food subsidies, but only because of the exceptionally favorable balance-of-payments situation. As is well known, these abundant sources of foreign and public revenues have a limited viability: oil exports are expected to last only a few years, massive U.S. foreign aid is politically unreliable, and workers' remittances from the Gulf countries have already started to decline. In an instrumentalist sense, it is evident that extractivist policies toward staple foods and cotton can only continue because of dualism in production, whereby the politically powerful rich farmers are engaged in the production of other lucrative crops. Strong growth in these new crops, which reflect the international comparative advantages of Egypt, remains, however, blocked by inadequate marketing channels and paralyzing bureaucratic constraints.

Conclusion

I started with the observation that the role of the state has been and is pervasive in the determination of agricultural and food prices and that, consequently, these prices and the growth and welfare effects they cause must be understood by reference to a theory of the state. This role of the state must be looked at critically against the backdrop of an increasingly food-dependent Third World with extensive malnutrition and rural poverty, and, in the United States, the joint occurrence of successful agricultural development and failing rural development (in terms of the defense of the family farm and rural communities). Why, then, do governments get so deeply involved in tampering with agricultural and food prices, and why are they so rarely able to engineer a policy package successful in terms of agricultural development, rural development, and satisfaction of the nutritional needs of their populations?

Several suggestions for answering these questions derive from the analysis in this paper:

1. One is that the broader the constituency of the dominance alliance, the lesser the instrumental functions of the state and the greater its relative autonomy. This, in turn, provides a greater possibility for the state to act in a capital-logic fashion to manage its accumulation and legitimation functions in a more balanced way. This, in a sense, is a reminder of the importance of more participatory forms of government

if the contradictions between growth and welfare implicit in the manipulation of prices are to be minimized.

Another is that developmentalist price policies toward agriculture have generally originated in strong instrumental demands from the farm lobbies. In the advanced countries, the decline in the political representation of farm interests and the increase in per capita incomes are undermining the political legitimacy of expensive farm budgets to support income parity. It is thus likely that farm policy will increasingly be made more in terms of economywide capital logic (cheap food, increasing agroexports) and less in terms of instrumental demands or rural development purposes.

2. The joint achievement of growth and farm income objectives through the use of price policy instruments, which is advocated under the purpose of rural development in the more unimodal farm production structures, is possible only with relatively egalitarian initial patterns in the distribution of farm assets. The more unequal the distribution of farm assets, the more regressive the subsequent distribution of income in agriculture, and the more agricultural development is obtained at the cost of rural underdevelopment or the expulsion of rural populations. Use of the price instrument for the joint achievement of agricultural and rural development thus requires an egalitarian land settlement or land reform policy and either the subsequent control of social differentiation or a strong demand for labor in the rest of the economic system.

3. The structural feature of social disarticulation is a major determinant of the capital logic for cheap labor and cheap food which, in turn, has been singled out by Schultz and others as the key factor responsible for stagnation of production and rural poverty. It is important to recall here that social disarticulation is not a mere stage of growth that precedes the turning point in the Kuznets U curve in the distribution of income. It is defined by a particular social class structure and is backed by usually nondemocratic forms of government. The question of distortions in agricultural price policy and the logic of the state in undervaluing food prices must consequently be seen in terms of the urgent need to transform the social basis of disarticulation.

4. In evolving from generally developmentalist regimes to more bureaucratic and authoritarian forms of government, many Third World countries have sacrificed planning for the sake of short-run stabilization policies. The result has been an increasing myopia in the handling of price policy, the dominance of short-run income distribution objectives over long-run growth promoting goals, and the primacy of immediate crisis response over the considerations of long-run food security. To a large extent, the continuing agricultural and food difficulties of many

Third World countries are consequently nothing more than the epiphenomena of the underlying crises of their states.

Acknowledgments

I am indebted to K. Subbarao and J. J. Dethier for useful critiques and suggestions.

Notes

1. By agricultural development, I mean output and productivity growth; rural development refers to improvements in the level and distribution of income in the sector (which include, in particular, issues of employment, land concentration and landlessness, and on- and off-farm sources of income).

2. I am referring here to the repeal of the Corn Laws in 1846, which allowed the entry of cheap U.S. wheat into England, collapsing the domestic price of wheat and eliminating landlords' rent. It was made famous by Ricardo's advocacy of comparative advantages and free trade, the so-called Manchester Doctrine.

References

Alderman, H. "Egypt's Food Subsidy and Rationing System: A Description." International Food Policy Research Institute, Washington, D.C., 1982. Mimeographed.

Andrews, M., and Alain de Janvry. "Intersectoral Terms of Trade, Income Distribution and Growth." Paper presented at the meeting of the International Association of Agricultural Economists, Jakarata, Indonesia, August 1982.

Bale, M. D., and E. Lutz. "Price Distortions in Agriculture and Their Effects: An International Comparison." Staff Working Paper No. 359. Washington, D.C.: World Bank, October 1979.

Baran, P., and P. Sweezy. *Monopoly Capital.* New York: Monthly Review Press, 1966.

Bardhan, P. "Authoritarianism and Democracy: First Anniversary of New Regime." *Econ. and Polit. Weekly* 13,11 (March 1978): 529–532.

Cochrane, W. *The Development of American Agriculture: A Historical Analysis.* Minneapolis: University of Minnesota Press, 1979.

Collier, D., ed. *The New Authoritarianism in Latin America.* Princeton, N.J.: Princeton University Press, 1979.

Cuddihy, W. *Agricultural Price Management in Egypt.* Staff Working Paper No. 388. Washington, D.C.: World Bank, April 1980.

de Janvry, Alain. "The Role of Land Reform in Economic Development: Policies and Politics." *Am. J. Agri. Econ.* 63(1981):384–392.

de Janvry, Alain, and E. Sadoulet. "Social Articulation as a Condition for Equitable Growth." *J. Dev. Econ.,* forthcoming.

de Janvry, Alain, and K. Subbarao. "Agricultural Price Policy and Income Distribution: A Computable General Equilibrium Model for India." Paper presented at the American Agricultural Economics Association meeting, Logan, Utah, August 1982.

Gardner, B. *The Governing of Agriculture.* Lawrence: Regents Press of Kansas, 1981.

George, P. S. *Aspects of the Structure of Consumer Foodgrain Demand in India, 1961/62 to 1973/74.* Res. Rep. No. 12. Washington, D.C.: IFPRI, November 1979.

Hadwiger, D., and R. Talbot. "The United States: A Unique Development Model." In *Food, Politics, and Agricultural Development: Case Studies in the Public Policy of Rural Modernization,* ed. R. Hopkins, D. Puchala, and R. Talbot. Boulder, Colo.: Westview Press, 1979.

Healy, D. "Development Policy: New Thinking About an Interpretation." *J. Econ. Lit.* 10(1972):757–797.

Hirschman, A. *The Strategy of Economic Development.* New Haven: Yale University Press, 1958.

―――――. *Essays in Trespassing: Economics to Politics and Beyond.* Cambridge: Cambridge University Press, 1981.

Kalmanovitz, S. *Desarrollo de la Agricultura en Colombia.* Bogota: Editorial la Carreta, 1978.

Krishna, Raj. "India's Sticky Economy and Some Noneconomics of Nonperformance." Sarabhai Lecture, Ahmedabad, 1981.

Krishna, Raj, and G. Ray Chaudhuri. "Some Aspects of Wheat and Rice Price Policy in India." Staff Working Paper No. 381. Washington, D.C.: World Bank, April 1980.

Kuznets, S. "Economic Growth and the Contribution of Agriculture: Notes on Measurement." In *Agriculture in Economic Development,* ed. C. Eicher and L. Witt. New York: McGraw-Hill, 1964.

Mellor, J. "Food Price Policy and Income Distribution in Low-Income Countries." *Econ. Dev. & Cult. Change* 27(1978):1–26.

Mitra, A. *Terms of Trade and Class Relations.* London: Frank Cass, 1977.

O'Connor, J. *The Fiscal Crisis of the State.* New York: St. Martin's Press, 1973.

Olson, M. *The Logic of Collective Action; Public Goods and the Theory of Groups.* Cambridge, Mass.: Harvard University Press, 1965.

Owen, W. "The Double Development Squeeze on Agriculture." *Am. Econ. Rev.* 56(1966):43–70.

Patnaik, U. "Contribution to the Output and Marketable Surplus of Agricultural Products by Cultivating Groups in India, 1960–61." *Econ. and Polit. Weekly* 10(1975):A-90, A-100.

Peterson, W. "International Farm Prices and the Social Cost of Cheap Food Policies." *Am. J. Agri. Econ.* 61(1979):12–21.

Poulantzas, N. *Political Power and Social Classes.* London: New Left Books, 1975.

Radwan, S., and E. Lee. "The State and Agrarian Change: A Case Study of Egypt, 1952–77." *Agrarian Systems and Rural Development,* ed. D. Ghai, A. R. Khan, E. Lee, and S. Radwan. London: Macmillan Press, 1979.

Rey, P. P. *Les Alliances de Classes.* Paris: Maspéro, 1976.

Saleh, A. "Disincentives to Agricultural Production in Developing Countries: A Policy Survey." *Foreign Agriculture* 13 (Supplement, 1975):1–10.

Sanders, T. "Food Policy Decision-Making in Colombia." American Universities Field Staff, Report No. 50/1980, South America, Hanover, NH.

Scobie, G., and R. Posada. "The Impact of Technical Change on Income Distribution: The Case of Rice in Colombia." *Am. J. Agri. Econ.* 60 (1978):85–92.

Schultz, T. W. "Constraints on Agricultural Production." In *Distortions of Agricultural Incentives,* ed. T. W. Schultz. Bloomington: Indiana University Press, 1978.

Schultze, C. *The Distribution of Farm Subsidies: Who Gets the Benefits?* Washington, D.C.: Brookings Institution, 1971.

Singleman, J., and E. Wright. "Proletarianization in Advanced Capitalist Economies." Center for Class Analysis, University of Wisconsin, Madison, 1978. Mimeographed.

Subbarao, K. "Food Production, Prices, and Income Distribution: A Review of Recent Trends in India." Paper presented at the Conference on Independent India: The First Thirty-Five Years, University of Santa Cruz, Calif., June 1982.

Thurow, L. *The Zero-Sum Society: Distribution and the Possibilities for Economic Change.* New York: Basic Books, 1980.

Vogeler, I. *The Myth of the Family Farm.* Boulder, Colo.: Westview Press, 1981.

Discussion

Robert Paarlberg
Harvard Center for International Affairs
Cambridge, Massachusetts

The question addressed by Professors Bates and de Janvry, the question of why governments act as they do, has been considered at length. In my comments on these two papers, I will try to move the discussion to a parallel concern: how governments might be persuaded to act differently. Specifically, how can the governments of nonindustrial countries be persuaded to alter those food and agricultural policies now in place that tend to undervalue agricultural commodities? How can these governments be persuaded to provide more favorable terms of trade to their own agricultural sectors?

If this is our concern, we have reason to be attracted to one part of Professor de Janvry's argument in particular. It was not his intent to pass favorable judgment on any of the four country cases reviewed, but one of those four cases—that of India—does suggest one political logic that might lead at least some nonindustrial countries to provide more generous terms of trade to agriculture, a political logic that might influence states to eventually escape from the trap of "urban-biased" agricultural policies.[1]

Professor de Janvry argues that the Indian government has at times intervened in an exceptional way to overvalue food and agricultural commodities. It has intervened to improve the terms of trade available to its own agricultural sector. The motive of the Indian government for doing so—as described by Professor de Janvry—is admittedly self-serving. The emerging capitalist class, still too weak to confront the peasantry and the working class, uses the overvaluation of agricultural commodities to "buy" political support in rural areas—especially from those landlords and large commercial farmers that receive the major share of benefits from this policy. Whatever the motive at work, we are presented with at least one circumstance in which food and agricultural policies in a major nonindustrial country, departing from the norm, have been relatively more generous to at least some rural producer interests.

Taking this exceptional case as a point of departure, does India's experience provide any grounds for hope that governments in other nonindustrial countries can also find ways to escape from the syndrome of an urban-biased agricultural policy? I would answer "yes," but not for the reasons that Professor de Janvry might recognize. It must be recognized that India's more generous food pricing policies—to the degree that they are more generous—have derived from a political logic quite different from that described by Professor de Janvry.

De Janvry has chosen to highlight the relatively generous agricultural price policies noted in India since 1965. His argument would not be able to account for the much less generous agricultural price policies noted in India before 1965. Particularly during India's Second Five Year Plan (1956/1957–1960/1961), the agricultural sector was not favored with advantageous terms of trade. It was during this Second Plan period that agriculture's share of public sector outlays fell quite sharply, from 37.2 percent of the total in the First Plan, to only 23.1 percent in the Second Plan. It was during this Second Plan period that the Government of India signed its first long-term agreement with the United States to accept progressively larger shipments of P.L. 480 wheat on concessional terms, which depressed producer prices. From 1959–1963, Indian wheat prices actually declined, in nominal terms, at a time that the general price index was trending upward. The result was such a disincentive to Indian wheat producers that the increase in acreage under wheat and the growth of Indian wheat production stopped. The failure to provide incentive prices during this period was later acknowledged by responsible Indian officials: The farmer "was not getting an economic or incentive price and this was one of the main constraints which impeded any further private investments in agriculture. . . . The absence of substantial investment in agriculture explained why there was general stagnation in agricultural production."[2]

Indian agricultural policies during this period were geared to meet the short-run demands of urban consumers (and urban industrial planners) at the expense of agricultural producers, large and small. The result, by the middle 1960s, was lagging production and rapidly growing consumption, sustainable only through a further expansion of P.L. 480 imports. This pattern of urban bias, found in Indian agricultural policy before 1965, is not unlike that which Professor Bates described so well in his discussion of contemporary food policies in Africa.

As Professor de Janvry implies, India's more recent experience has been rather different. Price incentives provided to commercial farmers by the government of India improved significantly in the middle 1960s. But Professor de Janvry's analysis of Indian food policy provides a misleading explanation for this change. The decision to improve price incentives was not, at the outset, an effort on the part of a weak emerging capitalist class to buy some short-run political support in the countryside, from landlords and commercial farmers. As Professor Bates might suspect, support from farmers did not yet count for much among those jockeying for political influence in the capital city. Farmers in India remained politically weak, at least from the vantage point of New Delhi. It was not to secure political support from rural groups, but to better meet the immediate needs of urban groups, that India's political leadership revamped its food and agricultural policies in the middle 1960s.

By 1964–1965, the policy of denying the agricultural sector in India had reached something of a natural limit. The ungenerous terms of trade that had been offered to Indian agriculture had become such a burden on India's own production performance as to finally pose a clear and present danger to the short-run welfare of urban consumers. Indian production had fallen so far behind internal demand that free-market prices finally stopped their downward slide, and began a sharp upturn. Not even a continued expansion of P.L. 480 imports was enough to moderate prices. In 1964, despite exceptionally favorable weather, which produced an above trend harvest, and despite much larger P.L. 480 imports, free-market prices paid by urban consumers began to increase at an annual rate above 20 percent.[3] It was in response to this adverse development, which directly threatened the objectives of a policy elite still oriented toward urban demands, that Indian agricultural policies at last underwent significant revision. The motive was not, at first, to buy support from rural-based power groups.

The changes that took place in Indian agriculture in 1964 and 1965 are not to be underestimated. In October 1964 a Foodgrains Policy Committee recommended a 15 percent increase in wheat procurement prices. In January 1965, this important offering of higher incentive prices to Indian wheat producers was then institutionalized through formation of the Agricultural Prices Commission. Later in 1965, Indian Agriculture Minister C. Subramaniam succeeded in pushing through the cabinet a significant package of new commitments to agriculture, including stepped-up production and import of fertilizer, plus the import and use of newly available seeds for high-yielding varieties of wheat. Agriculture's share of public sector outlays was also increased, back up to 31 percent of the total by the end of the decade.

These more generous policies did improve the terms of trade available to Indian agriculture, especially to large commercial wheat producers—

those best positioned to make use of the new Green Revolution technologies. Two years of very bad weather, in 1965 and 1966, delayed the evident payoff from these policy changes, but within four years after good weather returned, by 1971, Indian wheat production had actually doubled from the level attained in 1964.

As an interesting secondary effect, once the Indian government began to provide more generous terms of trade to agriculture, the distribution of power and influence within the Indian political system began to change as well. Commercial wheat producers were suddenly more prosperous, and because this newfound prosperity was so heavily dependent on public policy, these farmers sought more effective means to influence policy. Invigorated by their own rapid income growth, they became better organized, and began to use not only their political weight in key state governments, and their influence within the newly created Agricultural Prices Commission, but also to employ direct lobby action to press their case in New Delhi for still larger input subsidies and still higher procurement prices. An unintended result of India's more generous agricultural policies, then, was the strengthening of those rural power centers that had earlier been too weak to demand attractive government price and subsidy guarantees. The political weight of rural interests, which Professor de Janvry uses to explain the origins of India's more generous price policy, is perhaps better understood as one of the more interesting *consequences* of that policy.

For those who wish to be optimistic, perhaps India's experience can thus be a source of reassurance. At a point (perhaps at a point dangerously close to the stagnation of their own agricultural sector), elites in poor countries may find that the only way to continue to meet the escalating demands of their own urban consumers is to improve the terms of trade offered to rural food producers. Up to a point, urban-biased agricultural policies may be self-perpetuating. But beyond that point, by an internal logic, they may become self-correcting.

This observation leads me to the only point of difference that I might have with Professor Bates, who describes African agricultural policies as leaving little room for the possibility of change, let alone self-correction. Changes or corrections may be slow in coming; they may not be seen until many more years of needless inequity and inefficiency have been visited upon the long-suffering farmers of Africa. But the possibility (and the possible direction) of self-generating changes in agricultural policies in Africa would seem to merit some attention. Professor Bates is so thorough and so convincing in his discussion of why governments act as they do that we are almost left unable to imagine that they will ever be able to act differently. One appreciates, in this regard, the keen sensitivity to the process of change exhibited in the work of Professor de Janvry.

Professor de Janvry's paper is also noteworthy for its attempt to view the food and agricultural policies of the United States from within the same conceptual framework to bear on food policies in India, Colombia, and Egypt. The question posed in his discussion of U.S. food policy has been raised by others several times already during the course of this conference: Why is it that the United States (along with so many other wealthy industrial countries, such as the nations of Western Europe and Japan) so often intervenes to overvalue agricultural commodities, to ensure more generous terms of trade to agriculture? It goes against intuition that the United States, an urbanized industrial country, would single out its agricultural sector for such political favors (just as it goes against intuition, perhaps, that so many rural nonindustrial countries have selectively favored urban interests).

In searching for a way out of this puzzle, my own preference would be to stress similarities rather than differences between food policies in the United States and those in so many nonindustrial countries. In the United States no less than in the nonindustrial world, the short-run political demands of urban consumers for "cheap food" are dominant. But in the United States, as well as in Western European countries and Japan, the modernization of agriculture has sharply reduced the cost of producing food. Industrial growth has simultaneously increased personal income. In these countries, as a result, food will be relatively cheap for urban (and suburban) consumers no matter what the agricultural policies of the government might be. If food is relatively cheap to begin with, urban consumers will be less likely to object to modest governmental interventions that favor rural food producers. The budgetary cost of such interventions can be relatively small, since in these countries rural food producers are relatively few in number (and are becoming fewer in number with every passing year). Despite some overvaluation of agricultural products in these countries, continued agricultural productivity growth will ensure that the real price paid by consumers for food will not only remain low, but may even continue to fall.

Such are the blessings to political leaders of a modern and productive agricultural sector. Consumers in rich industrial countries, especially in the United States, find themselves so favored by the falling price of food that they scarcely notice when their government intervenes to slow the rate of that fall. Productive agriculture is a political godsend. It permits governments in industrial countries to pacify both producers and consumers at the same time. Those nonindustrial nations that are now seeking rapid industrial development without having first taken care to promote agricultural development run a political risk. They may reach a point, sooner or later, of satisfying the demands of neither producers nor consumers.

Notes

1. For an extended discussion of the problem of "urban bias" in poor countries, see Michael Lipton, *Why Poor People Stay Poor* (Cambridge, Mass.: Harvard University Press, 1977).

2. Quotation from C. Subramaniam, former Indian Minister of Agriculture. See Subramaniam, *The New Strategy in Indian Agriculture* (New Delhi: Vikas, 1979), p. 5.

3. R. N. Chopra, *Evolution of Food Policy in India* (Delhi: Macmillan India, 1981), p. 119.

Discussion _____

Bruce Gardner
Department of Agricultural and Resource Economics
University of Maryland

The topic addressed by de Janvry and Bates—why governments intervene in agricultural markets—is a topic that probably would not have been included in a discussion of agricultural markets ten years ago. What has changed? There has emerged a quite remarkable consensus of the left, center, and right that governmental intervention in agricultural commodity markets has had undesirable results in almost every instance, in every country. The authors of the two papers in question share this discouraging assessment. De Janvry asks "why are [governments] so rarely able to engineer a policy package successful in terms of agricultural development, rural development, and satisfaction of the nutritional needs of their populations?" (p. 208). Bates summarizes the central themes in policy formation in Africa as follows: "Governments engage in bureaucratic accumulation and act so as to enhance the wealth and power of those who derive their incomes from the public sector; they also act on behalf of private factions, be they social classes, military cliques, or ethnic groups. . . . economic redistribution [is] often from the poor to the rich and at the expense of economic growth" (p. 165). And he says that the evidence is sufficient to discredit any approach based on the conviction that governments are agencies of the public interest.

What do we mean when we assert that governmental policies have failed or been counterproductive? The point has a strongly normative sense: governments have not fostered growth, efficiency, or equity, goals which we believe governments should be aiming at; moreover, in so far as we can determine governments' motivations, they do not even seem to be very interested in achieving these things. Our response to this situation is typically simply to reassert what we think governments ought to be doing. The harder question, though, is to understand why governments behave as they do.

The papers by Bates and de Janvry make admirable attempts to address this question. They are thoughtful, intelligently argued, and well written, and make serious efforts to come to grips with some quite difficult issues. Yet in the end I think neither paper gets us very far toward a useful understanding of governmental behavior.

The main source of difficulty is that for the most part the level of explanation is too general. The form of explanation is: governments do the things they do because they are responding to the coalition of citizens best able to direct public activity to serve their private interests. Thus, we say that the milk support price is as high as it is because dairy farmers have been able to mobilize political support for their interests by campaign contributions and other means. This is not quite an empty hypothesis, but it does not bring us back far enough to observable causal variables.

Compare the economic theory of the individual consumer. We have a standard model: individuals maximize utility subject to budget constraints. Specifically, Mr. A bought this particular shopping cart full of groceries because by that set of purchases he was best able to use his funds to satisfy his wants. This is not quite an empty hypothesis either. But the useful part of demand theory is what helps us to identify observable characteristics of individuals or products, like prices, income, age, and so forth, that can be used empirically to explain quantities of goods purchased.

Analogously, a useful theory of policy would identify observable characteristics of interest groups or their environment that can be linked to political results. Both papers have partially baked components of such a theory, but their main interest is in the more general type of explanation.

I want next to discuss some details of the general theories of governmental action presented in the two papers. The two are quite different, which is striking because they are looking at basically the same phenomena. The conceptual machinery that our authors come equipped with appears to make a great difference, particularly in the set of propositions to which the authors assign subjective prior probabilities so near zero that they aren't even mentioned. Thus, the authors suggest quite different lists of hypotheses about governmental behavior for further testing. Background differences are evident in comparing the reference lists of the two papers. They cover a lot of the same ground and are each quite extensive, with 40 items in de Janvry and 67 items in Bates, yet there is only one common reference in the two lists.

Considering first de Janvry's ambitious paper, he begins with a nice overall statement of the issues and central questions to be answered in attempting to understand governmental behavior. Then he proceeds to tackle the issues using a set of concepts quite different and broader than those we are used to seeing in economics. He draws on a body of social theory expressed in terms of concepts derived from Marx, Weber, and several schools of their followers. This type of social thought tends not to be taken seriously, to put it mildly, by U.S. economists generally, and even less so by agricultural economists. But I think Marxian thought and its offshoots are particularly worth paying attention to in the theory of policy, because policy brings in institutions, constraints, and behavior quite different from what we are used to in standard economics, and to which the application of our standard tools is especially problematical.

As I read the general form of de Janvry's argument, the key elements are: (1) the fundamental role of *class,* (2) the importance of *contradictions* and their resulting *crises,* and (3) the idea of *capital logic* as a requirement of economic policy in a commercial society. The paper employs these concepts to embody the main virtue of what Bates in his paper calls the "pluralist" approach; namely, it "views public policy as a reflection of private interests." However, de Janvry places stronger prior restrictions on what groups constitute politically relevant interest groups by his focus on social classes. A more crucial analytical difference between pluralists and de Janvry appears when he brings in the notion of the power bloc as "the class alliance in control of the state." The subsequent discussion seems to me to imply the unstated but fundamental corollary that anybody outside the alliance has no power to influence policy.[1] It carries through to the first hypothesis listed in the conclusions, which says that the broader the power-wielding bloc in a government, the less damaging will be the government's intervention. Speaking of "power bloc" in the singular again means, I assume, that those who are not in the bloc have no power. There seems to be a maintained hypothesis throughout the paper that the state is inevitably a monopoly and the data to be explained involve who will control it.

The monopoly model of government seems to me implausible, even in nondemocratic countries. For one thing, as de Janvry says in his criticism of the Frankfurt school, the approach has no explanation for this "fundamental asymmetry in the power of different social groups." Also missing is sufficient recognition of constraints on the ability of governments to act and to maintain their positions. I thought at first that the concept of "capital logic" was to be de Janvry's vehicle for expressing such constraints, but it is not. Capital logic narrows down the requirements and needs of the ruling class, but it does not limit the ruling class from carrying out whatever policies it requires.

This becomes clear, relatively speaking, in the use de Janvry makes of the concept of "social disarticulation." His third hypothesis (p. 209) is that "social disarticulation is a major determinant of (the capital logic for) cheap labor and cheap food." Social disarticulation is not characterized by the sort of thing you might expect, such as Milton Friedman on a date with Jane Fonda. What it means is that the demand for consumption goods is demand out of the nonlabor income of the rich. This is said to require low wages, which in turn requires low food prices. Maintenance of this state of affairs over time, which is what de Janvry calls its capital logic, therefore requires policies that hold down food prices. This is not a constraint in the usual sense of conditions imposed by nature or others that limit the activity set of a maximizer. It is a conditional constraint of the form: I have to maintain my chickens because I need eggs.

The "capital logic" argument also has a serious weakness in economic theory: the hypothesis that "the level of monetary wages is principally determined by the level of food prices at low income levels" (p. 190). This presumably is a version of the "iron law of wages," which uses Malthusian and Ricardian ideas to infer a perfectly elastic long-run supply of labor at or near a subsistence wage. I had thought this theory to have been decisively refuted by the fact that its implication—that real wages cannot rise in the long run—fails to hold even in the poorest countries.

Returning to the paper's political theory, the notion of constraints imposed by people and events outside the ruling class does play a role in the discussion of contradictions and crises. This is my reading of de Janvry's fourth hypothesis, encapsulated in the final sentence of the paper, in which he says, "To a large extent, the continuing agricultural and food difficulties of many Third World countries are consequently nothing more than the epiphenomena of the underlying crises of their states" (pp. 209–210). This means that some policy moves are to be explained as a response to events beyond the control of the government (although perhaps determined by the situation that placed the governing class in power).

The danger at this point is that we can explain policies too easily. We explain policies under hypothesis three by saying that policy keeps the economy in the form the ruling class wants; but if the ruling class finds its economy falling apart, we explain its policies under hypothesis four as a response to that situation. And if the policy fails and the government falls, we get a new monopoly alliance in charge. We thus have no possible test of the ruling-class monopoly theory of policy.

This raises the bottom-line question for our purposes today: whether the general system of social thought that de Janvry employs can illuminate the causes of agricultural policy. Does it give us insights otherwise unattainable?

Even if the theory of a monopoly state is false, it might prove fruitful to take this theory as a maintained hypothesis and spell out its implications in terms of testable hypotheses. This is in fact what I interpret de Janvry to have done in the most interesting part of his work. Some empirically testable versions of his hypotheses on pages 208–209 are:

1. Deadweight losses are smaller, the larger the relevant interest group.
2. Deadweight losses are smaller, the more equal the prepolicy distribution of income between interest groups.
3. Policies will be more damaging to agricultural interests, the larger the share personal consumption expenditures above subsistence are accounted for by nonlabor income.

I am not enthusiastic about these hypotheses, but it would be interesting to see how they would fare in a systematic confrontation with the data. As stated, the hypotheses don't depend entirely on the monopoly theory of the state; (1) and (2) are quite consistent with Peltzman's pluralist theory in his 1976 *Journal of Law and Economics* article, or later work by Becker.[2] This essentially is a formalization of Stigler's earlier work on regulation, taking it from a quasi-monopolistic "capture" theory of regulation toward the pluralist position. I will allude to this literature (unmentioned by either author) because it seems more pertinent to our subject than any of the works cited in the papers of Bates or de Janvry.

Turning to the paper by Bates, he considers four alternative approaches to the explanation of governmental action. The approach to which he is most favorably disposed, the "pluralist" approach, essentially views governmental activity as expressing the outcome of conflict between private interests who vary in political power. Thus, the issue is: Who gains from governmental activity and who pays for these gains? The outcome reflects which groups have the most power. This seems to me also the single most accurate, simple picture of what goes on in governmental behavior, in my experience (not so much in developing countries but in the United States): to understand what is going on in the U.S. Congress we think of the various representatives as engaged in dividing a pie while also struggling over who is going to pay for baking it.

A second approach which Bates finds to be helpful is the idea that governments act so as to promote the well-being of those who are in the government. One sees this idea, too, applied not only to developing countries but also to the United States or other advanced countries, for example, in the literature on budget-maximizing bureaucrats. This view I find rather less convincing for the United States, at least, for the same reasons that de Janvry's monopoly theory of the state is unsatisfying. Both the dominant-bureaucracy and class-alliance monopoly hypotheses are particularly hard to reconcile with the existence of democratic institutions. One has to counter the argument that any rents politicians or bureaucrats have managed to generate for themselves can be bid away by opposition candidates or parties who can promise to throw the rascals out, and from time to time may be called upon to do so. The result is that the expected rate of return from becoming a politician or bureaucrat should be roughly the same as the expected returns from engaging in another profession requiring similar skills. Of course, both public-choice theorists on the right and Marxists on the left have developed many models in which democratic governments do not yield effective political competition. And if the government has monopoly power, then rents may be obtained. But even if democratic institutions are absent or ineffective in preventing this monopoly power, there remain constraints on its exercise, notably unrest, revolutions, and other costly situations. As Bates says, such threats mean that regimes must mitigate opposition, and probably the cheapest way of doing so is by channeling part of the wealth that government can generate to the groups in which opposition is most likely to arise. But when we talk of government distributing rewards and costs among people outside the government, whether to win votes or avoid revolution, we are back in the first approach, seeing governmental activities as primarily redistributive in a pluralist context where even those outside the governing coalition have political power.

Bates is hard on his other two approaches. One of these, derived from radical political economy, he says, is associated with "dependency theory." His criticisms of this approach seem to me appropriate, but the particular hypotheses that he attacks are not the most plausible ones that come out of radical political economy or the Marxist tradition.

Bates is also very critical of the idea that governments act to promote the public interest, characterizing the idea as naive, wrong, and occasionally pernicious. He calls this "the political theory of welfare economics," but it is more appropriate to say that standard welfare economics does not have any political theory. Welfare economics is a quasi-normative discipline that does not purport to say anything about why governments do the things they do.

When economists speak on practical policy issues, it is true that many of us have been guilty of too easily assuming that if we could tell governments what they ought to do in the area, say, of promoting agricultural research, or education, or price stabilization, then the natural next step would be for the government to go ahead and undertake these recommended activities. Nonetheless, there is a strong tradition among both right-wing and left-wing economists to deny the public interest view. I think myself that this rejection of the idea of government seeking to promote the general good has gone too far. I agree that in the United States, what Congress is most interested in doing is dividing up the pie, but this does not mean that Congress is indifferent to questions about the size of the pie. The approach I prefer is to look at government as maximizing the society's joint income subject to redistributional demands, inevitably costly to satisfy, that reflect the differential political power of various interest groups.

In sum, while I disagree with Bates on some of his four approaches, I agree that the most plausible way to look at governmental activity is as a redistributional struggle among interest groups, this as against de Janvry's view of a class alliance that gains a monopoly of the state. The pluralist analogy is that the industrial organization of the state is equivalent to monopolistic competition or is a cartel with free entry.

In any case, the bottom line for present purposes is how to explain which groups do best in the political arena. My main problem with Bates' discussion is its generality; he expects some pressure groups to win out politically, but has no systematic way of predicting which ones. He does discuss one hypothesis (seemingly accepted as obvious) that I think is mistaken; namely, he seriously overstates the importance of numbers of people in a pressure group in promoting the interests of that group. On page 172 he says, "Clearly, were competitive elections contested by rival parties in Africa, agricultural policy could *not* be so strongly biased against rural dwellers. With less than 10 percent of their populations in cities, most nations would contain electoral majorities composed of farm families; and electoral incentives would almost inevitably lead politicians to advocate pro-agrarian platforms in their efforts to secure votes and win power."

Compare the U.S. situation, where elections are contested by rival parties, and where the farm sector is much less than 10 percent of the population, yet the positions that win out are quite pro-agrarian in the sense that they distribute substantial sums per capita to the farm sector from the nonfarm sector. Clearly, a large fraction of the population in one's interest group is neither a necessary nor sufficient condition for that group to do well politically. (From my point of view a more satisfactory discussion of interest group size is in Peltzman's paper cited above.)

Bates' paper has the substantial virtue of actually reporting some systematic empirical work, although I didn't think it was integrated well with the theory. He finds that the ideology of political leaders matters and that policies at times do appear aimed at maximizing GNP. Ideology seems often to consist in large part in the acceptance of a general socioeconomic theory of some kind, so it isn't surprising that differences in ideology lead to different policies; but whether the differences in policy reflect *normative* differences isn't clear.

I have been able to fulfill the discussant's duty to find things to disagree with in these papers, but I want to close by expressing my appreciation for the seriousness and intensity of thought in both of them. It would also be churlish not to recognize the lack of any other literature that contains the testable hypotheses and empirical tests of them that I criticize these papers for not containing.

Notes

1. There could be a revolution, but this would only transfer ownership of the state to another class alliance.

2. S. Peltzman, "Toward a More General Theory of Regulation," *Journal of Law and Economics* 19(1976): 211–240; G. Becker, "A Theory of Political Behavior," CSES Working Paper No. 006. (Chicago: University of Chicago, September 1981).

Part 5 ⎯⎯⎯⎯⎯⎯⎯⎯⎯⎯⎯⎯⎯⎯⎯⎯⎯⎯

7

The Role of Trade in Food Security and Agricultural Development

Robert L. Thompson
Department of Agricultural Economics
Purdue University

In exploring the role of markets in the world food economy, previous papers have focused mainly on internal, developing-country issues. Professors Poleman and Srinivasan have discussed the nature and extent of hunger, its causes and possible solutions. This is the heart of the food security problem. Dr. Reca has discussed the types of price policies employed, and has analyzed their effects on food production and consumption in developing countries. Professors Bates and de Janvry have treated the question of why governments do what they do to their agriculture.

Professor Johnson launched the discussion with a review of recent trends in the supply, demand, and international trade in food and other agricultural commodities. He documented the recent growth in agricultural trade, particularly in grains, and argued that a "world food system" capable of providing food security to most peoples has now evolved. In the present paper we turn our attention to the role international trade can and should play in achieving food security and agricultural development goals in the developing countries.

I will first discuss the concept of food security and the alternative means of achieving it, including international trade. Agricultural development is identified as an important contributor to food security. The discussion then turns to the sources of agricultural development, the structural transformation an economy undergoes in the course of economic development, and the manner in which trade may facilitate or impede this process. The interrelationships between the sectors and the need for balanced growth are emphasized. I then treat the recent international trade performance of the developing countries and argue that those following an export-oriented growth strategy have performed relatively better than other countries. Nevertheless, there are numerous barriers to developing country exports, and world markets, particularly for agricultural commodities, are often highly volatile. Together with a presumed decline in the terms of trade of primary exporters, this leads many developing countries to doubt the advisability of relying on agricultural trade as a source of either food security or agricultural development. The merits of these arguments are evaluated, and the paper ends with implications for agricultural trade, development, and food security policy in developing countries.

Food Security

Food security is one of the most fundamental objectives of every nation. A great deal has been written about "food security" since the "world food crisis" of 1973-1974; however, so many different definitions of the term have been used by different authors that one quickly becomes bewildered. Therefore, before taking up the role trade may play in achieving food security, it is incumbent upon us to clarify how the term "food security" is being used here.

Individual Food Security

Ultimately, food security concerns the individual or the family unit. Its principal determinant is purchasing power–income adjusted for the cost of what that income must buy. The rich do not suffer from food insecurity. It is a phenomenon of the poor, who spend the largest fraction of their income on food. The greatest incidence of absolute poverty, and hence of food insecurity, is in the rural areas of developing countries—among the smallest farmers and particularly the landless rural laborers. The other group most susceptible to food insecurity is the urban poor—particularly unskilled laborers, who are often recent migrants from rural areas. Food insecurity can result from a drop in income, as from loss of a laborer's employment or drop in a farmer's crop yields, or from a sharp increase in the price of food that reduces the purchasing power of that income and hence the amount of food it can buy.

Economic growth and development are concerned with raising the per capita income and well-being of a country's population. As incomes grow, the fraction spent on food declines, and the chances of falling into food insecurity drop. It is important to emphasize that for this to happen, development must be broad-based and reach the lowest-income members of society. It is possible for national average income to grow without raising the incomes of the poorest people or reducing the threat of food insecurity. If trade can contribute to broadly-based economic development that increases the employment possibilities and per capita incomes of the poorest members of society, it can play an important role in reducing food insecurity. We will return to this issue at a later point in this paper.

Aggregate Food Security

Much of the recent literature on food security concerns what we might call a country's "*aggregate* food security," not the *individual* food security discussed so far. One definition of aggregate food security is ensuring adequate food supplies to feed the country's population at reasonable prices, regardless of how crop yields fluctuate from year to year. As Professor Srinivasan emphasized in his paper, objective nutritional criteria for defining aggregate food security are very difficult, if not impossible, to define. As a result, many analysts have used as a criterion a given fraction of trend consumption. This physical availability of food is a necessary, but not a sufficient condition, for individual food security, as we saw above. The latter requires individual purchasing power as well.

In any given year, due to adverse weather conditions or a pest infestation, crop yields, and therefore total food output, may fall. Each agroclimatic zone has a different probability of this occurring, with this risk being greater in semiarid and arid zones that lack irrigation potential. A country's trend food consumption may be maintained in bad crop years by drawing down stocks, reducing exports, or increasing imports. If the country lacks reserve stocks, does not export food, and its balance-of-payments situation does not permit larger food imports, national consumption must fall. The internal price of food will rise to ration the available supply, either officially or via a black market, causing food insecurity for the poorest households.

It is not only individual purchasing power that determines food security. Unless adequate reserve stocks are held within the country, national purchasing power is an important determinant of national food security. Analogous to the individual case, national purchasing power is the amount of foreign exchange available to pay for food imports, adjusted for the price that has to be paid. As Professor Johnson illustrated in his paper, world agricultural commodity prices vary greatly from year to year. Valdés and Konandreas (Valdés 1981, pp. 25–53) have shown that in recent decades, 25 percent of the variability in developing country food import bills is accounted for by price variability and 75 percent by quantity variability. A sudden increase in food import price, a drop in export revenue, or both can make it difficult, if not impossible, for a developing country to meet target consumption levels. Therefore, still a third criterion for food security has been proposed. Siamwalla and Valdés (1980, p. 26) suggest using "the deviation from trend in foreign exchange earnings minus the excess expenditure over the trend in food imports for the current year." Here, we identify two aspects of the role trade can play in assuring food security. The country must be able to generate foreign exchange revenue from exports if it is to import the food needed to ensure food security when its crops fail.

Self-Sufficiency

Some observers add a further twist to their definition of national food security. They add the notion that a country should produce enough food itself to provide some minimum level of food intake per person to protect against the contingency that it might be unable to import food at any cost—as in time of war or embargo. (Japan's and Switzerland's farm policies have this as one objective.) Less extreme versions of this definition emphasize the desirability of food production growing at least as fast as demand. They identify a decline in national food security with an increase in the deficit between domestic consumption and production. Some call this the long-term food security problem, and call the year-to-year variability around trend the short-term food security problem. Clearly, domestic agricultural development is necessary to increase the rate of growth in domestic food output.

Food Security Strategies

Before concluding our discussion of food security, we need to discuss the three categories of proposals for improving food security in developing countries: reduce fluctuations in crop yields (around an increasing trend if possible), maintain reserve stocks, and import any shortfalls. Crop yields may be increased through a package of higher-yielding inputs, including improved seeds, fertilizer, and pest and water control. The latter two, plus timely availability of purchased inputs, can also reduce yield fluctuations.

A great many proposals have been advanced since 1974 for creating national reserve stocks. Those studies that have carefully evaluated the costs of maintaining cereal stocks in tropical countries have concluded that for each country to maintain its own stocks would be a very expensive means of achieving food security. For example, World Bank estimates for landlocked Sahelian countries show that the facilities' depreciation, interest on the capital tied up in inventories, physical losses, and costs of handling would total about $150 per ton per year (cited in Yudelman 1982). This does not take into account the additional costs from having to roll over the stocks every year or so to prevent the grain from going out of condition. (The same study reports that the Canadian Wheat Board charges its contract customers $10 per ton per year for the same services.) The cost to a developing country of the foregone investments that could be made with the resources tied up in idle grain reserves can be quite high.

Most analysts have concluded that when domestic production falls short of target consumption, imports are a cheaper means of ensuring food security than reserve stocks. This may involve diverting foreign exchange away from other purposes on which national economic development depends, such as imports of technology, training, machinery, raw materials, and intermediate inputs not available in the country. As a result, national economic development, employment, and income growth may slow down. Instead, the country might borrow foreign exchange to import food. This has the disadvantage of using up the country's limited line of credit in nonproductive investment.

There are several alternatives, however. First, the country might receive food aid. Many developing countries are now skeptical of relying on food aid for their food security. In 1974, the year in the last decade when food aid was most needed, the quantity available dropped substantially. Food aid appropriations are made in national currencies, so when the price of grain goes up, as it did in 1973-1974, the quantity available for food aid declines. Moreover, bilateral food aid usually has political strings attached. Nevertheless, an International Emergency Food Reserve of 500,000 metric tons was established in 1976, and contributions of 588,000 m. tons were made to it in 1981. A new Food Aid Convention, negotiated in 1980, raised the minimum annual contribution of food aid from 4.2 to 7.6 million tons.

An alternative to physically storing grain, the expense of which has been amply demonstrated, is storing foreign exchange. In effect, a country could maintain an interest-bearing bank account in foreign exchange as a reserve to be drawn upon to import grain in years of crop shortfall. This would avoid the high cost of facilities, handling, and losses from physically storing grain. Although the account would earn interest, the country would forego the development inputs the reserved foreign exchange could otherwise buy. Konandreas, Huddleston, and Ramangkura (1978) proposed an insurance scheme through which, in exchange for annual premium payments, a developing country could stabilize cereal consumption at a relatively stable cost.

In 1981 the International Monetary Fund (IMF) established a food facility as an extension of its Compensatory Fund Facility to provide financial assistance to offset fluctuations in countries' food import bills, either because of shortfalls in domestic production or higher world food prices. Although untested so far by any major crisis, the International Emergency Food Reserve, the Food Aid Convention, and the IMF Food Facility together should substantially reduce the cost, perceived or real, to developing countries from relying on imports to maintain national food security in short crop years.

Agricultural Development and Structural Transformation

In the previous section, we saw that poverty is the most important cause of individual food insecurity. Broadly-based economic development that raises the per capita incomes of all strata of the population, then, can make an important contribution to combating food insecurity. Successful general economic development requires that the agricultural and nonagricultural sectors of an economy be developed in a balanced fashion. The experience of more and more countries is demonstrating that if either lags too far behind, bottlenecks arise that tend to derail national economic development. As we shall see, international trade plays a strategic role in this process.

The Structural Transformation

In the early stages of development the agricultural sector bulks large relative to the rest of the economy—both as an employer of labor and in the fraction of the GNP it produces. In low-income countries it is not uncommon to find 70 percent of the population employed in agriculture, with the agricultural sector producing 30 to 50 percent of the GNP. In the course of economic development, however, a structural transformation occurs as agriculture's role in both employment and GNP generation declines. In middle-income countries the fraction of the population employed in agriculture falls to around 40 percent and agricultural production to 10 to 20 percent of the output. In high-income countries only about 6 percent of the employment and 4 percent of the output come from agriculture.

The World Bank (1982, p. 39) has characterized agricultural development as "successfully transforming agriculture for the benefit of the economy and of the people who live and work in agriculture—and will eventually leave it." Two forces drive the structural transformation in which many people move from farm to nonfarm employment. The reason they leave agriculture is found in the numbers cited in the last paragraph. In every case cited a smaller percent of the population produces a larger proportion of the output in the nonfarm sector than in agriculture. In other words, output per worker and therefore incomes per capita, are higher in the nonfarm sector than in agriculture. This, together with the fact that no developing country provides as good schools, medical services, and amenities to its rural residents as are available to its cities, provides the incentive for people to move. In the process, national income per capita rises because a large fraction of the work force becomes employed in higher productivity employment.

The second force driving the structural transformation comes from the demand side. As one's per capita income rises, the mix of products one purchases undergoes a significant change. Consumption of manufactured goods and services rises faster than demand for farm products. Moreover, of the income spent on "food," the fraction that actually goes to the farmer grows more slowly than that consumed in processing, transport, packaging, and other services. On both counts, the fraction of income spent on farm products falls. To satisfy this changing structure of demand as incomes grow, the nonfarm sector has to grow faster than agriculture.

The mix of agricultural products demanded also changes as incomes rise. Demand for meats, dairy products, fruits and vegetables, and edible oils grows faster than demand for staple cereals and tubers. There also seems to be a hierarchy of preferences for carbohydrate sources moving from tubers to coarse grains to rice and finally wheat.

The shifts in demand to some extent reflect tastes and preferences. Some shifts are also motivated by the different amounts of time required to prepare different foods for the table. As the value of the housewife's time rises, and especially if she takes a job outside the household, foods requiring less time to prepare assume a larger role in the diet. This is one reason why the Japanese housewife, for example, is serving more and more bread instead of boiling rice. Not only does the balance between agriculture and the rest of the economy change as incomes rise, but the mix of products produced within agriculture itself must change as well.

The dynamic changes in the structure of supply and demand in a developing economy mean that the mix of products it imports and exports must change. Before addressing the role of trade more specifically, we need to examine the changes that occur in the production sectors of the economy during development. The production possibilities of the underdeveloped economy are determined by its labor force, which is mainly unskilled, its endowment of land, the given climatic conditions, and whatever mineral wealth may lie below the land. Technology is usually at a fairly rudimentary level.

Sources of Growth

The main sources of growth in output from both agriculture and the rest of the economy are technological improvements, increasing use of purchased intermediate inputs, and capital investment, whether in machinery, education, infrastructure, or research and development. These factors make the existing land and labor more productive. Growth in productivity, rather than in bringing more land and labor into production, accounts for most of the output growth we observe—whether in the United States or in low-income countries.

For example, Hayami and Ruttan (1971) have shown that three categories of factors each account for about one-third of the differences we observe across countries in agricultural productivity: (1) size of land area and labor force; (2) past investments in rural education and agricultural research; and (3) use of "modern inputs" such as fertilizer and tractors. Until the last decade most growth in farm output in developing countries came from bringing more land into production, but this has changed of late. A growing share of their output growth now comes from "nontraditional inputs."

The above delineation of the sources of economic growth provides one clue as to why growth needs to be balanced between agriculture and the rest of the economy if economic development is to succeed. The nascent manufacturing sector is constrained initially by the size of the domestic market (defined as the number of people times their purchasing power). (Exports of manufactures are not generally feasible until the industry matures.) Agricultural development requires increasing quantities of tools, machinery, fertilizer, transportation, commercial services, and construction. As rural income grows, farm families are able to buy more consumer goods, such as textiles and clothing, cookware and dishes, radios, kerosene, processed foods, vegetable oils, bicycles, and home construction materials. Many of these farm inputs and consumer goods can be produced by the nascent manufacturing sector. Broad-based agricultural development, which increases rural incomes of all classes, can exert a powerful stimulus to development of the nonfarm sector. The increased income this generates in the nonfarm sector has a further multiplier effect, as it is in turn spent on other goods and services, including farm products. Without this stimulus from rural expenditures, nonfarm development and general economic development tend to lag.

The Role of Trade

Some of the inputs needed for successful development of both sectors are not or cannot be produced within an underdeveloped country. No country contains deposits of all the minerals needed for producing the various fertilizers and metals. Most countries produce less petroleum than they consume. Whatever essential inputs a country lacks must be imported. Other essentials for development may be producible in the country at a later stage, but must be initially imported by an under-developed country to get the process moving. This frequently includes machinery and equipment, technology, and technical skills. The latter may be acquired by sending nationals overseas for training, which in effect means importing their skills when they return. Some consumer goods not available locally will also be imported from overseas.

Abstracting from this last point, we see that imports can play a very strategic role in supplying many of the high-payoff inputs needed to make agricultural and nonagricultural development successful. For them to play this role, however, the developing country must have the foreign currency or international purchasing power needed to pay for them. This can come from only five sources: (1) export earnings, including remittances from "exported" labor; (2) borrowing; (3) foreign investment; (4) selling assets to foreigners, e.g., land or mineral rights; or (5) foreign aid. For various reasons, it is essential for a developing country to come up with much of the needed foreign exchange through exporting.

Although foreign aid can help substantially, it often has more political strings attached than most developing countries are comfortable with. Most donors link much of their aid to purchases of goods and services from the donor country, which is not always what the recipient wants or needs.

Selling assets to foreigners is not attractive as it means, in effect, giving up part of the country's birthright or sovereignty. Foreign investment can certainly help, particularly since modern technologies, management skills, and training often come in the deal. Nevertheless, no self-respecting country wants all of its modern manufacturing sector to be owned by foreigners. In addition, foreign investors are not in the charity business. They only invest where they expect to be able to make a market rate of return in the form of repatriable profits. Over the long run, the investor expects to take out more foreign exchange than he brought in. Finally, many developing countries fear the power of large foreign firms whose annual turnover may exceed the GNP of the country in which they are investing.

Borrowing foreign exchange from foreign commercial banks, foreign governments, or from development banks is possible up to a point. A country's creditworthiness is based in part on its past export performance, so underdeveloped countries may find themselves with limited borrowing potential.

Need for Agricultural Exports

Developing countries must export to earn the foreign exchange needed to import essential inputs for development. Since agriculture is the largest sector of such countries, exports will almost certainly have to come from there, at least initially, unless a country possesses exportable mineral wealth. If the agricultural sector fails to produce a positive balance on agricultural trade, it may well put a brake on national economic development.

Agricultural production, then, must not lag behind the food demand of the country's population. This is not to argue that countries should be self-sufficient in every food product. That would be ridiculous, given the differing agroclimatic conditions required by different crops. The point is merely that in the early stages of development, if a mineral-poor country's agriculture cannot generate an export surplus, its economic development will likely be retarded.

Later, once the manufacturing sector gets moving and matures sufficiently to start exporting, the export pressure on agriculture may ease. As incomes rise and the product mix demanded changes, other agricultural products or inputs may be imported. That has occurred in a number of middle-income countries, for example, whose demand for meat is growing rapidly. The modern broiler production technology is easily transferable among countries, and broiler output has grown much faster than production of feed grains and high protein feeds. As a result, many middle-income countries have become large importers of feed grains and soybean meal, purchased with receipts from exporting manufactured products. It is important to recognize that there is no reason whatsoever for agricultural trade to balance in any country at any state in its development. Rather, it is a question of in which products a country has a comparative advantage at its given stage of development. This changes through time, as the structure of a country's supply and demand changes through development.

Dynamic Changes in Comparative Advantage

A primitive economy's comparative advantage is determined by its endowments of unskilled labor, often undeveloped land, and its technologies, which likely lag behind those used in more developed economies. Its capital stock—whether in physical capital, infrastructure, education, or research—is usually small relative to other countries. The high-payoff sources of economic growth that alter the production structure of the economy are the same forces that alter its comparative advantage and trade patterns. In the long run its comparative advantage may change dramatically from that determined by its original primary factor endowment.

The dynamic force determining how comparative advantage changes stems from the decisions on how scarce investment resources[1] are allocated among sectors and among projects within sectors. There is an infinite range of possibilities for how the limited funds could be spent. It is incumbent upon the investment decisionmakers—whether they be public or private—to choose those projects that will yield the highest social rate of return and, in turn, will do the most to further the process of economic development.

Other things being equal, projects that make sparing use of the country's relatively scarce, hence dear, factors of production, and make intensive use of its more abundant (cheaper) resources will have higher rates of return. For example, in a labor-abundant landscarce country, such as many in southern and eastern Asia, investments that save land and use labor, such as breeding higher-yielding varieties, building irrigation systems, or increasing fertilizer use, will have a higher rate of return than, say, tractor mechanization. The opposite situation prevailed in the United States and Canada in their early stages of development. One must not be too dogmatic on this, however, since relative factor scarcities may change in the course of development. For example, Japan's rapid industrialization in the past two decades has pulled so much labor out of agriculture that investments in laborsaving mechanization have become socially profitable. After the U.S. frontier closed, land became relatively more scarce, and yield improvements were made through hybridization and increased fertilization.

Some industries are characterized by greater ease of substitution between inputs. At least until recently, it has been easier to substitute capital for labor in agriculture than in industry. As a result, in Asia, where labor is relatively cheap, we find the capital:labor ratio to be much lower in agriculture than in industry. In the United States, where labor is relatively expensive, the capital:labor ratio is higher in agriculture than in industry. (With the increasing use of robots in industry, this may be changing now.) Such reversals in the input use intensities in the same industries in different countries mean that primary factor endowments may not be a reliable guide to a country's comparative advantage.

Moreover, in a mature economy, as we have seen, the structure of production possibilities and hence comparative advantage may bear little resemblance to its original factor endowments. Rather, the accumulated pattern of past investments as reflected in its existing capital stock is the main determinant of comparative advantage. With the international mobility of capital, technology, and to a lesser extent labor, a country can generate a comparative advantage in new industries through an appropriate investment strategy.

Ultimately, the pattern of comparative advantage may be determined by: unique land resources, including both soil and mineral deposits; distance and thus freight rates to foreign markets, particularly for bulky commodities with high transport costs per unit of value; the size of the country market (population times income), which affects whether economies of scale are realized in manufacturing industries; and political stability, which determines the size of the risk discount attached to foreign investment, and therefore the capital mobility to the country.

Impediments to Agricultural Development

The pattern of private investment decisions made within a country, which determines the rate of its economic development and of future comparative advantage, is strongly influenced by the relative prices of products as well as by the public investment decisions of the government. For example, in many developing countries the governments follow what may be called a "cheap food policy," in which the price of food is depressed to artificially low levels relative to nonfarm prices (see Chapter 4). This is often implemented through a system of administered price ceilings. Food aid accepted and dumped on the local market also depresses food prices. An overvalued exchange rate may further contribute to this because it implicitly taxes all exports and subsidizes all imports. Industry benefits from lower prices for both food and the imported inputs it needs to develop.

As food is the largest component of consumer expenditures, by holding the price of food down, both industry and government can hold down the wages of their employees. This raises returns in industry, lowers the government payroll, and, in turn, taxes needs. The net result is that agricultural returns are reduced due to the depressed ratio of prices received to prices paid. This lowers the rate of return to investments in productivity-increasing improvements, and agriculture tends to stagnate.

The private incentive to improve agriculture is further reduced by the proclivity of governments to allocate most (often as much as 90 percent) of the investible resources under their control to the nonfarm sector and urban areas. Rural areas receive little investment in transportation, communications, health services, schools, or other social services. Little goes to agricultural research to develop improved technologies adapted to the conditions of the country. Governments also tend to create high-cost parastatal marketing monopolies, which further reduce the price received by farmers. Finally, import substitution industrialization strategies create high-cost industries that can only survive under stringent import tariff or quota protection. This raises the price of both inputs and consumer products purchased by the farm family. The overall effect is to retard agricultural development.

The lag in agricultural development slows down nonfarm development in two ways. Because farm incomes do not increase as rapidly as they could, demand for domestically produced manufactures grows more slowly. Second, if food production fails to expand apace with demand (particularly from the growing urban sector), this will put upward pressure on the price of food as long as foreign exchange constraints preclude satisfying this growing demand through imports. Wages must then rise, lowering industrial returns and increasing the amount of taxes the government must collect to meet its payroll. On the other hand, if food prices are kept down by increasing imports of food, this uses up the scarce foreign exchange so badly needed for imports essential for development, or for debt service on past borrowing from abroad. On either count, the rate of economic development is depressed.

Agricultural development will not occur if price policies are biased against it at every turn, and if government invests little or nothing in it. Krishna (1982) has argued that, as a rule of thumb, perhaps 30 percent of investment should initially go to agricultural development, although this may be lowered at a later stage in development. Once cost-reducing technologies are adopted in agriculture, and if supply expands at a sufficiently rapid pace, this will put downward pressure on the price of agricultural relative to nonfarm products. This provides the same benefit to industrial development that many governments have tried to artificially force prematurely. The difference is that with technological change, unit costs decline. Net farm income does not necessarily decline just because the prices received for farm produce decline. This, together with the overvalued dollar, caused the "farm problem" in the United States in the 1960s.

With an equilibrium exchange rate and no other market interventions, the terms of trade need not turn against agriculture when output expands in this manner. Rather, the excess can be exported at the world market price, generating additional foreign exchange revenue. If effect, what happens in this case is that technological change gives the country a greater comparative advantage in those products than it had before.

The Role of the Exchange Rate

The dynamic changes that occur in comparative advantage and trade during the development of an economy generally cause improvements in the country's balance of payments and, in turn, in its exchange rate, other things being equal. As the country's currency appreciates relative to foreign currencies, the price of imports in domestic currency falls and the competition for less efficient domestic industries increases. The exchange rate appreciation makes all the country's exports dearer to foreign purchasers. This causes problems for export products in which the country previously had only a marginal comparative advantage. At any time one could rank all goods from greatest comparative advantage to greatest comparative disadvantage. The list changes constantly as capital investments and technical change alter the goods included, their rankings, and the distance between them. Changes in the exchange rate shift the boundary between exportables and importables up and down in order to keep international payments in balance.

A country may realise a sudden increase in export revenue by discovery of mineral wealth or through a significant increase in the price of an export, such as occurred in the Organization of Petroleum Exporting Countries in the past decade. This so strengthened their exchange rates that other exports and sectors competing with imports suffered. Venezuela is a good case in point. On the other hand, countries with a heavy petroleum import dependence saw their balance of payments deteriorate and their exchange rates depreciate. This made exporting easier, made more goods exportable, and increased the protection of domestic industries from import competition.

As a country's industrial sector grows and matures to the point that it becomes a successful exporter, this may reduce the competitiveness of its farm exports in the world market. This places increased importance on agricultural research to maintain a competitive position.

The statement that a country has nothing to export can be true only if its exchange rate is artificially set so unreasonably high that it is priced out of the foreign markets for all its potentially exportable goods. The answer to exporting in such a case is to devalue the currency. What this does to the price of imports and, in turn, to the cost of living may not, however, be politically attractive to its policymakers. The challenge is to not just treat the symptoms by artificially or cosmetically altering prices, but rather to treat the disease itself. This means making the necessary investments to increase the country's comparative advantage in certain goods relative to other potential exporters, so that the more politically desirable exchange rate becomes the market equilibrium rate.

Recent Experience in World Trade

The previous section provided a stylized discussion of the role trade could play in development, based upon the experiences of many developing countries—both successful and unsuccessful. It focused upon the internal workings of individual countries and the structural transe of development. It abstracted completely from trends in world trade and the developing countries' role therein. The *World Development Report, 1982* provides a very useful analysis of these developments, from which the following review is synthesized.

From 1955 to 1980, world trade expanded five times, at an annual rate of 5.1 percent. With the rapid growth in the world economy from 1955 to 1973, world trade grew at 8 percent per year. This growth was dominated by trade in manufactures, fuel, and minerals. During this period agricultural trade grew by only 4 percent per year. As a result of their slower growth rate, agricultural exports fell from 60 percent to 30 percent of the total value of world exports.

With the slowdown in the world economy after 1973, the rate of growth in world trade slowed as well. Since 1973 world trade has grown at only 4.7 percent per year. The growth in agricultural trade has accelerated, however, to 4.8 percent per year. In the 1970s the volume of world agricultural trade grew 45 percent, while world farm output grew only 24 percent, thereby raising the share that is traded. Taking a longer perspective, the share of world grain production entering trade tripled from 6 percent in 1950 to 18 percent in 1980.

As farm trade has grown, a significant shift in the commodity composition of agricultural trade has transpired. The fastest growing commodities in the 1960s and 1970s were feed grains, high-protein meals, vegetable oils, and nonfat dry milk. Sugar, tropical beverages, and cotton had low rates of growth, and beef was intermediate. Trade in other fibers declined.

For developing countries as a group, agriculture products accounted for 30 percent of their total export revenue in the late 1970s, compared to 60 percent in 1955. Nevertheless, some low-income countries still depend on one or a few agricultural exports for the bulk of their foreign exchange revenue.

In some countries a large fraction of total farm output is exported. FAO data for 1979 for 90 developing countries show that more than half of farm output is exported in 10 countries, more than 20 percent in 30 countries, and more than 10 percent in 50 countries (FAO 1981). Agriculture earns a significant fraction of the total foreign exchange revenue in many developing countries. Yet the economies of those with the largest export exposure are often destabilized by the variation in world demand and price. Export diversification would be desirable for them.

In the past 30 years, many developing countries have taken advantage of the growth in world trade to expand their exports. Several developing countries experienced substantial successes in increasing crop exports, including the Ivory Coast, Malaysia, the Philippines, Thailand, and

Brazil. But the real success story concerns the developing countries' entry into manufactured exports. In the 1970s their exports as a whole grew at 5.2 percent per year, but their manufactured exports grew at 14 percent per year. Their market share for manufactured imports into industrial countries rose from 7 percent to 13 percent in the decade of the 1970s. The share going to other developing countries shrank. At first, exports were concentrated in clothing, textiles, footwear, and other labor-intensive goods. Recently, the more experienced countries whose wage rates have risen as they have moved into middle-income (or newly industrializing) status, have shifted into goods requiring more design, higher technology, and more skilled labor and management. These include electronics and other consumer goods, and both light and heavy engineering. South Korea and Taiwan provide the much-heralded success stories in this category.

Only one-quarter of developing country exports are bought by other developing countries; the balance goes to the rest of the world. (Exports to other developing countries tend to be more capital-intensive than those to high-income countries.) Industrial country imports from the developing countries grew at only 1.1 percent per year from 1965 to 1978. Although on a smaller base, imports from them by the centrally planned, and particularly the oil-exporting countries, grew much more rapidly.

Although the industrial countries have been, at least until recently, fairly receptive to increasing their manufactured imports from developing countries, they are extremely protectionist when it comes to agricultural imports. They do buy half of the low income countries' agricultural exports, although these tend to be concentrated in noncompetitive products like tropical beverages and fruits.

From the mid-1950s to the late 1970s, world grain trade grew from about 30 to 230 million tons per year. Most of the increase originated in North America and the European Community, and to a smaller extent from developing countries, particularly Brazil and Thailand. The growth in imports occurred mainly in the centrally planned economies, the middle-income developing countries, and the oil-exporting countries. The USSR and Eastern Europe alone accounted for one-third of the total increase. To a preponderant extent, these importers were experiencing rapid growth in incomes and thus in meat consumption. They lack a comparative advantage to increase food production to match the rate of demand growth. This effect is reinforced in the centrally planned economies by cheap food policies that result in consumers paying much less than the world price for meat and other livestock products.

Despite the numerous trend projections in the wake of the "world food crisis" of 1973-1974, very little of the growth in grain trade over the past decade has been to meet the food needs of low-income countries. As Professor Johnson showed (see Chapter 1), in the developing countries as a group food production has increased faster than population. In sub-Saharan Africa, however, per capita food production has declined for more than a decade. These economies suffer severe balance-of-payment constraints on their ability to commercially import cereals. Although their food aid imports have risen, a somewhat better performance of agriculture in other parts of the world has tended to offset Africa's increased imports. China has enjoyed a number of agricultural development successes, but it is relying increasingly on imports to meet the growth in urban cereals demand and for livestock feeding.

From this review of recent developments in world agricultural trade, we can conclude that the growth observed in the past decade has been fueled mainly by growth in per capita income and urbanization, rather than by growth in population. The middle-income importers' export successes have provided the international purchasing power for them to buy agricultural imports. The same applies to the oil exporters. The Soviet Union has exported gold among other products. Eastern Europe has paid for some of its food imports by borrowing abroad (export credits).

The evidence is very strong that economic growth has been faster in those developing countries that followed a balanced development strategy (in which potential export industries were promoted) than in those that followed an import substitution industrialization strategy and neglected their agricultural development. The most striking successes have been in eastern Asia and Southeast Asia, Latin America, and the Middle Eastern oil exporters. Many low-income countries have not achieved sufficient industrial development to begin relying on manufactured exports as a source of growth. Delays in agricultural development have held back economic growth in some countries, particularly in sub-Saharan Africa. The problems of food insecurity tend to be concentrated in these countries, and among poor people left behind by economic growth in countries experiencing faster growth.

Barriers to Realizing Trade Potential

Recently, as the industrialized countries have slipped into recession, their rate of economic growth and their volume of imports have declined. The shrinking import demand has put downward pressure on the prices of traded goods, particularly of primary commodities. The decline in both prices and quantities traded has depressed export earnings everywhere, but this has hit the developing countries particularly hard. As their ability to import and to service their foreign debt has declined, their rates of economic growth, which held up well through the 1970s, have dropped.

Despite the recent spectacular export successes of several developing countries, and the positive correlation between export performance and economic growth, many developing countries remain unconvinced of the gains they purportedly might achieve through larger exports and imports, whether of agricultural or nonagricultural products.

Declining Terms of Trade

Many reject the argument presented above, that early in its development a developing country that lacks exportable mineral wealth needs to expand agricultural exports to earn foreign exchange for development projects. They adhere to the notion of a secular tendency for the terms of trade of primary producers to decline. By implication, it would take more and more agricultural exports to import the same amount of manufactured goods. This argument is reinforced by the notion that to be agricultural is to be backward or undeveloped, and that a modern economy is industrial.

The declining terms of trade argument has its roots in the observations that demand for manufactures grows relatively faster than per capita income and demand for primary products grows relatively slower. The tendency to substitute synthetics for nonfood agricultural products such as fibers, it is argued, reinforces this tendency. The fallacy of this argument against exporting farm products is that it focuses only on the demand side of the equation. It ignores the dynamic changes that occur in the production structure of a developing economy as a result of capital accumulation. This investment includes, of course, cost-reducing technological change. If agriculture undergoes technological change that lowers its unit cost of production at least as fast as the unit price is falling, net incomes and rural welfare need not decline. Just because the terms of trade may move "against" a country, this does not necessarily lower its well-being. Its well-being is more likely to decline if its investments in agricultural research do not keep up with those in the rest of the world, since these are likely to be the source of the original decline in the terms of trade.

The empirical evidence is mixed on whether or not developing countries' terms of trade have declined. One can show almost anything one wishes by choosing the appropriate mix of products and the appropriate beginning and ending years of the period analyzed. Professor Johnson provides evidence in his paper (see Chapter 1) that the prices of several of the most traded agricultural commodities have tended to decline during the twentieth century relative to the prices of all other goods. Whether the countries exporting those products have become worse off as a result depends on how fast technological change has lowered their production costs.

As we saw above, the structural transformation of an economy and the accompanying changes in its comparative advantage are normal parts of economic development. In an agrarian economy that through economic growth becomes a high-income country, the relative importance of manufactures in its exports will increase. This has occurred in such "agricultural countries" as Denmark, New Zealand, and Canada. The important point is that an underdeveloped country may well have no other source of exports than its agricultural products. With some investment and at least a neutral policy environment, agriculture can help generate the foreign exchange needed to develop the rest of the economy as well as itself. If successful, the result sought by the terms-of-trade pessimists will occur as a natural consequence of economic development.

Finally, one needs to be cautious of the fallacy of composition. By lumping all primary commodities together, one discounts their heterogeneous character. Demand for some, such as livestock products and the feed grains used to produce them, grows rapidly as incomes grow. Demand for others, such as fibers, grows much more slowly. A few developing countries have seen their agricultural export receipts grow rapidly as they have shifted resources into crops whose export demand is growing rapidly. Two recent examples are Brazil's exports of soybeans and frozen concentrated orange juice, and Thailand's exports of maize and tapioca pellets. Many others are exporting fresh fruits, vegetables, and cut flowers—often by air—to high-income country markets. Countries that have not attempted to diversify and whose sole agricultural exports are in the low growth category have suffered.

Protectionism

Other observers despair at the restrictions imposed by protectionist policies of importers on developing country exports. However, by the 1970s most tariff barriers to manufactured imports in high-income countries had been eliminated as a result of the various rounds of multilateral trade negotiations under the General Agreement on Tariffs and Trade (GATT). In fact, the developing countries themselves are much more protectionist against manufactured imports than the high-income countries are. As a result, the share of manufactured exports from developing countries bought by other developing countries dropped significantly during the 1970s.

Nontariff barriers to manufactured imports by the high-income countries have become more important, however. These are particularly damaging to trade because, with them, efficient exporters have no way to compete on the basis of price. These nontariff barriers include quotas, "voluntary restraints," orderly marketing arrangements, price-maintenance agreements, countervailing duties and safeguard procedures, antidumping actions, subsidies for suppliers' credits, and incentives to domestic production.

The recent recession and associated unemployment have aggravated protectionist tendencies in the high-income countries, particularly for textiles, clothing, footwear, electrical goods, automobiles, and steel. The threat that further restrictions may be imposed on these and other import goods is likely to reduce exports of these products from developing countries. Protectionist sentiments often run highest in labor-intensive industries, the same ones in which developing countries first develop a comparative advantage.

The high-income countries are much more protectionist on agricultural than on manufactured imports. Whereas developing countries tend to depress agricultural prices, most industrialized countries artificially raise farm prices, with the objective of raising farm incomes. As a result, they have to restrict imports to avoid supporting the whole world price structure. By artificially raising returns in agriculture, excessive capital investment occurs, and too much labor is held in the sector. Technological changes may be adopted more rapidly than they otherwise would be, and too intensive use of variable inputs occurs. All these factors push up the volume of production at the same time that the artificially high prices reduce consumption. If a product is imported, this reduces the volume of imports on both counts. If production exceeds consumption, stocks accumulate. The only way to dispose of the stocks is to dump them on the world market for whatever price they can fetch, or to subsidize exports sufficiently to make the high-cost domestic product competitive. If this occurs in sufficient volume, the world market price is depressed. This has clearly been the case for dairy products, and perhaps also for cereals more recently. To the extent that developing countries import part of their food supply, one could argue that such policies benefit the recipients, who don't have to spend as much foreign exchange for their imports. The lower import price means that the food costs of the domestic consumers also decline. On the other hand, this reduces returns to the importing country's farmers and may slow down its agricultural development.

We argued previously how important it is for developing countries to increase their exports of agricultural products early in their development because they have few other options. By reducing developing countries' agricultural export possibilities, such high-income country policies slow down their economic development. Valdés and Zeitz (1980) recently carried out a study of the effects of such protection on agricultural exports from developing countries in the mid-1970s. They concluded that a 50 percent reduction in OECD-country trade barriers would increase agricultural exports from 56 developing countries by about 11 percent or $3 billion (in 1977 dollars). One-third of the additional export revenue would go to sugar producers, and another third to exporters of meat, beverages, and tobacco. Almost 60 percent of the gains would go to Latin America, over 20 percent to Asia, and 10 percent each to sub-Saharan Africa, North Africa, and the Middle East.

Another feature of this protectionism is particularly pernicious to the possibilities of exporters' developing processing industries to add more value to the primary product prior to export. Importers tend to levy higher duties on processed goods than on the raw product. This "cascading" provides very high effective rates of protection to the value-adding process in the importing country. This thwarts another means by which developing countries might raise national income and export revenue.

World Price Instability

Many protective agricultural policies are of the nontariff form. The United States imposes quotas on beef and dairy product imports. The European Community charges a variable levy on maize imports. In Japan a government wheat import monopoly buys at the world market price and resells the wheat on the domestic market at a fixed administered price. In the Soviet Union prices are rigidly fixed at the same level for years at a time.

The important feature of such policies is that they cut any link between domestic prices of the relevant commodities and their prices in the world market. As a result, domestic producers and consumers do not share in the adjustment to either bumper crops or crop failures in the rest of the world. In a similar fashion such countries "export" all of their yield variability to the rest of the world, rather than asking their consumers to share in the adjustment to the changing crop size. As a result, world market prices are much more volatile than they would be if all trading countries adjusted their consumption.

D. Gale Johnson (1975) has estimated that as much as 40–50 percent of the international grain price variability in the 1970s was due to government policies, not to the acts of nature that caused the crop yields to vary from year to year. In a study of the world wheat market, Shei and Thompson (1977) have shown that as more countries cut the link between world and domestic prices, the world market price becomes more volatile in response to any yield variability. A policy of freer wheat trade by all nations would substantially reduce world wheat price fluctuations.

This is not to argue that there should be no price variability. As long as crop yields fluctuate due to weather, the market price must be able to adjust to clear the market. The point to be made here is that the large price swings observed in the last decade or so have been larger than necessary, having been amplified by nontariff barriers to agricultural trade.

If we lived in a world of perfect certainty in which all future prices and crop yields could be foreseen correctly, a country could specialize in that small set of goods in which it has greatest comparative advantage and reap large gains from trade. In the real world, however, prices vary a great deal, and can be predicted, at best, in a probability sense. It is not uncommon to find underdeveloped countries with one or a few commodities accounting for large fractions of both their GNP and export revenue. If the export prices of these are highly volatile, the whole economy—everything from GNP to employment to the inflation rate to tax revenue—is on the whip end of the world market prices for these commodities. This can be terribly destabilizing to a country's economic development. In this situation—and there are a number of cases like this—a country is clearly too specialized in one or a few export crops. Its strategy should be to diversify its exports, so that no one export can have such a destabilizing effect on the economy as a whole. If decisionmakers seek to avoid risk, they may prefer to have a lower average annual total export revenue if the year-to-year variability around that average is also lower.

International price risk is often cited as the rationale for developing countries' unwillingness to rely on food imports as a source of food security and for their desire to become self-sufficient in food. Some countries that appear to have a clear comparative advantage in a cash export crop and have imported substantial parts of the food supply are increasing taxes on the exports to make their production less attractive, and raising prices of food crops to make their production on the same land more attractive.

Senegal, an important peanut exporter that imports rice and wheat, has done just this. Jabara and Thompson (1980) have shown that, if its policymakers are risk averse, this may have been a very rational strategy for Senegal in the face of large price variability for both its export and import products. These situations argue for the use of an extended concept of comparative advantage that explicitly recognizes the subjective costs (assuming risk aversion) associated with international price instability. Uncertainty in world market prices reduces the subjective returns from exporting and raises the subjective cost of importing. This is not an argument for autarky, but only observes that in a risky world, the optimum degree of specialization and volume of trade are probably smaller than if prices were more predictable.

Within a given year, hedging is an effective, low-cost means of transferring price risk to others who are more willing to carry it. There is no organized means, however, by which a country can hedge the prices of its exports and imports five years in advance, for example. This is a relevant concern because most investments have a pay-back period of a number of years. Relative future prices determine the relative profitability of the alternative investments a country might make. The choice of investments is the most important factor in modifying the shape of the country's future production possibilities and of its comparative advantage.

In a related piece Jabara and Thompson (1982) argue that the optimal trade policy under international price uncertainty is to distort the domestic prices away from the international prices, raising the price of the imported relative to the exported goods, if policymakers wish to minimize risks. The size of the optimum distortion is larger, the greater the variability in world prices and the greater the risk aversion of the policymakers.

This optimal policy under uncertainty is shown to be a variable levy that stabilizes internal prices around the level expected to prevail in the world market in the long run. The internal target price should be set equal to the expected international terms of trade plus the subjective risk cost. The tariff actually charged on any given day equals the difference between this target price and the price of the day in the country's port. The actual tariff rate would vary from day to day. (This is the same policy now employed by the European Community on its grain imports.)

No one small country acting in isolation can affect the international terms of trade, but all developing countries acting simultaneously to implement food import substitution programs would move the terms of trade against their suppliers and thereby make them worse off. Moreover, if these countries cut the link between domestic and international prices as suggested here (e.g., by imposing a variable levy or equivalent quantitative restriction), greater price instability would be expected in international markets, thereby further increasing the optimum price distortion. This suggests that freer trade, commodity agreements, or other schemes designed to reduce world commodity price variability could discourage self-sufficiency schemes and in turn promote greater international trade among countries as a result of reducing international price uncertainty. Part of the costs of such a stabilization scheme borne by exporters might be compensated by increases in the volume and value of trade. As price uncertainty is reduced for small countries, the expected welfare maximizing trade flows should converge towards what one would expect under free trade. The recent creation of the IMF Food Facility may reduce the policymakers' risk aversion. If so, this may increase the volume of LDC imports.

One final observation must be added to this discussion of world price instability. There is evidence that agricultural prices are more flexible, and therefore more variable, than industrial prices. Some argue that agriculture has a competitive structure, while the structure of industry is more concentrated. They argue that industry has the power to prevent price declines, while agriculture does not. An alternative explanation concerns the duration of the contracts used for transactions in the respective industries. The shorter the contract, the more often the price of a good can change. It is argued that the contracts in agriculture are written for a much shorter period than those in industry. For whichever reason, if agricultural prices are more flexible in the short run, they will also be buffeted more by monetary changes in the economy. In times of monetary expansion they will tend to rise faster than industrial prices, and in times of monetary contractions, to decline proportionately more than prices in industry. This suggests that some of the instability we observe in agricultural prices is a reflection of unstable monetary policies followed by the country.

Self-Imposed Impediments

To conclude this section on barriers to developing countries' realizing the potential benefits from trade, we must add that domestic policies are probably as important as the policies of other countries. Most developing countries overvalue their exchange rates, thereby taxing exports and subsidizing imports. Many have invested excessively in industrialization to replace imports, only to find themselves at least as dependent on imports, this time for raw materials, spare parts, and intermediate inputs. Most have followed a cheap food policy and underinvested in agriculture early in development, so that agriculture could not generate export revenue in the early stages. Most developing countries are themselves more protectionist than the high-income countries. By this means they often deprive growing sectors of cheaper sources of intermediate inputs or machinery, preventing them from becoming competitive in the export market. These policies have all worked against the potential contribution that trade could make to economic development.

Summary and Conclusions

In recent decades, with the growth in world agricultural trade and the accompanying evolution of international marketing institutions, a world food system has developed. This system has contributed to the economic development and food security of low-income countries. If permitted to function freely, it could do much in this regard.

Food insecurity at its most basic level is a problem of poverty, not of crop failures. Only poor people go hungry. Food imports contribute to national food security by assuring adequate supplies. They do not solve the individual's problem unless that individual has money to buy more food. Trade contributes to food security mainly by facilitating faster agricultural and economic development, thereby increasing per capita food production and incomes.

Food imports are a cheaper means of achieving food security than national reserve stocks of grain in tropical countries. Many developing countries are loath to rely on imports for their food security because of international price variability and their fear that exporters might embargo shipments in a bad year. Food aid has been an unreliable source of food security. Commercial food imports compete for the scarce pool of foreign exchange available for investment projects, and excessive use of foreign exchange for food imports will slow down the rate of economic development, unless the country has exportable mineral wealth. It may be desirable, therefore, to increase agricultural output, at least as fast as food demand grows. The recent creation of an International Emergency Food Reserve, a Food Aid Convention, and the IMF Food Facility should help reduce the foreign exchange drain on poor countries who need to import to provide food security in years when crops fail or when export receipts decline.

Agricultural and general economic development are closely interlinked through the markets for food, foreign exchange, labor, capital, and manufactured goods. Early in the development process agriculture is the largest employer and producer in the economy. If it fails to increase food production at constant or declining prices, agriculture can effectively put a brake on the country's general economic development.

As it develops, an economy undergoes a structural transformation in which the nonfarm sector grows faster than the agricultural sector. This is driven by the faster rate of growth in demand for manufactures and services as incomes grow. Because labor productivity and incomes are lower in agriculture than in the rest of the economy, per capita income cannot reach its potential level unless the structural transformation occurs.

As it develops, agriculture needs increasing quantities of inputs produced by the nonfarm sector. As the nonfarm sector grows, it must pull labor out of agriculture. Investment capital is also needed. Because agriculture is the largest sector, it is the main domestic source of investable funds to develop both sectors. Unless agriculture increases its marketed surplus, urban food prices will rise, or food imports must increase. Finally, for the manufacturing sector to grow, agriculture must provide a market for its output, since manufactured exports are unlikely until the nascent industrial sector matures.

Foreign exchange is required by a developing country to import machinery, technology, training, and other inputs such as fertilizer needed to develop agriculture and to transform the production structure of the economy. Unless a country has exportable mineral wealth, agriculture, as the largest sector, is really the only potential source of export earnings until a manufacturing sector can grow and mature. Because foreign exchange is such a critical necessity for development, a country needs to invest sufficiently in its agriculture at an early stage to efficiently expand production. It can earn foreign exchange from exports and save foreign exchange by reducing imports, if this can be efficiently done through technological change. If it succeeds, and if the foreign exchange is allocated wisely, this can significantly accelerate the pace of national economic development.

No country can be self-sufficient in all commodities. Different farm products require too large a diversity of climatic conditions for this to be efficient. Furthermore, in bad crop years a country may well need imports to provide food security. This is clearly cheaper than maintaining reserve stocks of grain in tropical countries. Successful agricultural development should, however, reduce the frequency with which this occurs. Investments in irrigation and pest control can help stabilize yields and reduce the likelihood of food insecurity.

The investments a country makes in the course of its economic development change its production structure and comparative advantage. As the industrial sector matures and begins to export, the pressure on agriculture to generate foreign exchange may decline. The income growth that accompanies development shifts the mix of products people demand toward more livestock products, fruits and vegetables, and edible oils. Many people change their principal carbohydrate source. Products for which demand is growing will need to be either produced in the country or imported. In some cases, the technology for producing the final product is more easily introduced than the technology for producing the necessary inputs. For example, many middle-income countries are producing broilers efficiently with imported feed grains and protein meals.

Whereas a country's agriculture needs to generate a balance-of-trade surplus early in development, at a later stage there may be a deficit on agricultural trade. If so, this merely reflects the changes that have occurred in the country's structure of production and consumption and in its comparative advantage. Agriculture is much too heterogeneous a sector for a sectoral balance of trade to have much meaning in a mature economy. What matters is the country's overall balance of payments, not that of any given sector.

The more foreign exchange earned by those industries in which a country has the greatest comparative advantage (other factors being equal), the stronger will be its exchange rate. As a result, less efficient industries will shift from the net export to the net import column. Discovery of previously unknown mineral wealth or a large increase in the world price of an existing export will have the same effect.

With the rapid postwar growth in international trade, many developing countries have successfully expanded their exports, and the share of these exports represented by manufactured products has grown significantly. Several countries have also diversified their agricultural exports into those commodities for which demand is growing fastest.

Since 1973, growth in manufacturing trade has slowed down, while growth in agricultural trade has accelerated. The latter growth has been fueled mainly by higher incomes, with middle-income developing countries, the oil exporters, and the centrally planned economies accounting for most of the growth. The poor developing economies, despite lagging agricultural production particularly in sub-Saharan Africa, have not contributed much to the growth in global imports. Contrary to many projections, population growth has not caused food imports to increase. Purchasing power is the principal factor.

As trade has grown, a number of structural shifts have occurred. A growing fraction of developing country manufactured exports has been bought by high-income countries. Tariff protection on manufactured goods imports has dropped to negligible levels in the industrial countries, although nontariff barriers are a growing impediment. This is particularly true for labor-intensive manufactured products in which low- to middle-income developing countries have a comparative advantage.

Developing countries themselves impose much greater barriers to manufactured imports than do the high-income countries. In addition, their own policies have often reduced their trade performance. Many overvalue their currencies, follow cheap food policies, invest too little in developing their agriculture, and prematurely promote import substitution industrialization. Those countries that have followed an export promotion development strategy have been most successful. They have not neglected their agricultural development in the process, however, and they recognize that it is cheaper to import some goods than to produce them domestically. Excess protection of inefficient domestic industries also puts a severe brake on development.

High-income countries are much more protectionist with respect to agricultural imports—especially of sugar, beef, tobacco, and beverages. This is particularly deleterious to low-income countries that lack alternatives to farm exports to earn the foreign exchange needed for development projects. Moreover, many of the protectionist measures are nontariff barriers to imports. These effectively keep out imports from more efficient producers. They also cause greater instability in world commodity prices. Perhaps half of the price variability in the affected commodities is caused in this manner, rather than by crop yield variability.

The instability and unpredictability of world commodity prices reduces developing countries' confidence in the gains from trade and in the ability of trade to provide food security. If policymakers are risk averse, they may prefer a more diversified economy with a lower average income as long as the variance around that average is also lower. Large variability in world prices and greater caution by policymakers may provide an incentive to increase the country's self-sufficiency in imported food, even though the cost of production may be higher than the cost of importing. To the extent that this occurs, the food exporting countries of the world have a vested interest in reducing the volatility of commodity prices.

The deterioration in the terms of trade of agricultural exporters has reduced many developing countries' interest in promoting agricultural exports. This, however, need not diminish the country's well-being. Net incomes may actually increase in the face of declining prices if cost-reducing technological change is being adopted. Relative price trends, in and of themselves, are an inadequate guide to producers' well-being. Moreover, relative price changes are an integral part of a dynamic economy undergoing technological improvements. It is inevitable that some sectors will progress faster than others.

Economic growth in the middle- and high-income countries has been the principal engine of growth for developing country exports. In the recent global recession the volume of international trade declined, which depressed the developing countries' growth rates. Their prospects for faster growth in the coming decade are closely linked to how fast the world economy rebounds from this recession. If Professor Johnson's prognosis is correct, that we are in for a period of slower income (and population) growth in the coming decade, this bodes poorly for developing country growth prospects.

We now have a world food system capable of providing food security to most peoples. If permitted to function freely, it would contribute as well to agricultural and general economic development. This too would reduce food insecurity. Two factors impede this. High-income countries' import policies impede developing countries from exporting products in which they have a comparative advantage. This applies to both agricultural and nonagricultural goods. In addition, the domestic agricultural policies and trade policies of many developed countries "export" their instability to the world market, increasing the variability of world prices. This reduces the desirability of relying on trade for food security and as an engine of growth. This is costly in terms of foregone income to importers and exporters alike.

Acknowledgments

The research on which this paper is based was supported in part by USDA cooperative research agreement no. 58-3J22-1-0325X and by a Title XII Strengthening Grant from USAID to Purdue University.

Notes

1. The available investable funds come from domestic savings—forced as well as voluntary—and foreign investors. One form of forced savings is tax revenue collected by the government, which is in turn invested in projects such as publicly owned industries and infrastructure and biological research in agriculture. The social rate of investment return exceeds what a private investor, who cannot internalize all the benefits, would receive. Note that because of its preponderant size in the economy, agriculture must also provide savings for investment in the rest of the economy during development. The challenge for a government is to tap the agriculture surplus through taxes or other means without depressing the rate of agricultural development.

References

Abbott, P. C. "Agricultural Trade and Economic Development: Implications of the Foreign Exchange Gap to 1990." Foreign Agriculture Economics Report, Economic Research Service. Washington, D.C.: USDA, 1983.

Abbott, P. C., and F. D. McCarthy. "Welfare Effects of Tied Food Aid." *J. Dev. Econ.* 11(1982):63–79.

Anderson, K. "Changing Comparative Advantage in Agriculture: Theory and Pacific Basin Experience." *J. Rural Dev.* 3(1980):213–234.

Balassa, R. "Exports and Economic Growth: Further Evidence." *J. Dev. Econ.* 5(1978):181–189.

Bhagwati, J. *Foreign Trade Regimes and Economic Development: Anatomy and Consequences of Exchange Control Regimes.* Cambridge, Mass.: Ballinger Publishing Co. for National Bureau of Economic Research, 1978.

Bruno, M. "Optimal Patterns of Trade and Development." *Rev. Econ. & Stat.* 47(1967):545–554.

Chenery, H. B. *Structural Change and Development Policy.* New York: Oxford University Press for World Bank, 1979.

Chenery, H. B., and M. Syrquin. *Patterns of Development 1950–1970.* London: Oxford University Press, 1975.

Falcon, W. "Food Self-Sufficiency: Lessons from Asia." In *International Food Policy Issues: A Proceedings*, pp. 13–20. Foreign Agri. Econ. Report No. 143. Washington, D.C.: ESCS/USDA, January 1978.

Flanders, M. J. "Agriculture Versus Industry in Development Policy: The Planner's Dilemma Reexamined." *J. Dev. St.* 5(1968):171–189.

Food and Agriculture Organization (FAO). *The State of Food and Agriculture.* Rome: FAO, 1981.

Goreux, L. M. *Compensatory Financing Facility.* Pamphlet Series No. 34. Washington, D.C.: International Monetary Fund, 1980.

Hathaway, D. E. "Food Issues in North-South Relations." *World Economy* 3(1981):447–459.

Havrylyshyn, O., and M. Wolf. "Trade Among Developing Countries: Theory, Policy Issues, and Principal Trends." Staff Working Paper No. 479. Washington, D.C.: World Bank, 1981.

Hayami, Y., and V. W. Ruttan. *Agricultural Development: An International Perspective.* Baltimore: Johns Hopkins University Press, 1971.

Hughes, H., and J. Waelbroeck. "Can Developing Country Exports Keep Growing in the 1980s?" *World Economy* 4(1981):127–148.

Jabara, C. L. "Terms of Trade for Developing Countries—A Commodity and Regional Analysis." Foreign Agri. Econ. Report No. 161. Washington, D.C.: ESCS/USDA, November 1980.

Jabara, C. L., and R. L. Thompson. "Agricultural Comparative Advantage Under International Price Uncertainty." *Amer. J. Agri. Econ.* 62 (1980):188–198.

———."The Optimum Tariff for a Small Country Under International Price Uncertainty." *Oxford Econ. Papers* 34(1982):326–331.

Johnson, D. Gale. "Increased Stability of Grain Supplies in Developing Countries: Optimal Carryovers and Insurance." *World Dev.* 4(1976):977–987.

———."World Agriculture, Commodity Policy, and Price Variability." *Am. J. Agri. Econ.* 57(1975):823–828.

Johnston, B. F., and P. Kilby. *Agriculture and Structural Transformation.* New York: Oxford University Press, 1975.

Johnston, B. F., and J. W. Mellor. "The Role of Agriculture in Economic Development." *Am. Econ. Rev.* 51(1961):566–593.

Keesing, D. B., "Trade Policy for Developing Countries." Staff Working Paper No. 353. Washington, D.C.: World Bank, 1979.

Konandreas, P., B. Huddleston, and V. Ramangkura. *Food Security: An Insurance Approach.* Research Report No. 4. Washington, D.C.: IFPRI, 1978.

Krishna, Raj. "Agricultural Growth, Price Policy and Equity." Paper presented at Conference of Agricultural Officers, World Bank, Washington, D.C., January 1982.

Krueger, A. O. *Foreign Trade Regimes and Economic Development: Liberalization Attempts and Consequences.* Cambridge, Mass.: Ballinger Publishing Co. for National Bureau of Economic Research, 1978.

———."Growth, Distortions, and Patterns of Trade Among Countries." Princeton Studies in International Economics No. 40. Princeton, N.J.: Princeton University, 1977.

_____."Trade Policy as an Input to Development." *Am. Econ. Rev.* (Pap. & Proc.) 70(1980):288–292.

Lewis, W. A. "Elmhirst Memorial Lecture: Development Strategy in a Limping World Economy." In *Rural Change: The Challenge for Agricultural Economists,* ed. G. Johnson and A. Maunder, pp. 12–26. Westmead, U.K.: Gower, 1981.

Little, I.M.D., T. Scitovsky, and M. Scott. *Industry and Trade in Some Developing Countries: A Comparative Study.* London: Oxford University Press for Organization for Economic Cooperation and Development, 1970.

McIntire, J. *Food Security in the Sahel: Variable Import Levy, Grain Reserves, and Foreign Exchange Assistance.* Research Report No. 26. Washington, D.C.: IFPRI, September 1981.

McKinnon, R. I. "Foreign Trade Regimes and Economic Development: A Review Article." *J. Int. Econ.* 9(1979):429–452.

Michaely, M. "Exports and Growth: An Empirical Investigation." *J. Dev. Econ.* 4(1977):49–53.

Reutlinger, S. "Food Insecurity: Magnitude and Remedies." *World Dev.* 6(1978):797–811.

Sarris, A. H. "Grain Imports and Food Security in an Unstable International Market." *J. Dev. Econ.* 7(1980):489–504.

Shei, S. Y., and R. L. Thompson. "The Impact of Trade Restrictions on Price Stability in the World Wheat Market." *Am. J. Agr. Econ.* 59(1977):628–638.

Siamwalla, A., and A. Valdés. "Food Insecurity in Developing Countries." *Food Policy* 5(1980):258–272.

Thompson, R. L., and P. C. Abbott. "On the Dynamics of Agricultural Comparative Advantage." Paper presented at USDA-Universities Consortium for International Agricultural Trade Research Conference, Bridgeton, Mo., 24–25 June 1982.

Tyers, R., and A. Chisholm. "Agricultural Policies in Industrialized and Developing Countries and International Food Security." Paper presented at 26th annual conference of the Australian Agricultural Economics Society, Melbourne, 9–11 February 1982.

Underwood, J. M. *Food Security and Food Policy in a World of Uncertainty.* Working Paper. New York: Rockefeller Foundation, November 1979.

Valdés, A., ed. *Food Security for Developing Countries.* Boulder, Colo.: Westview Press, 1981.

Valdés, A., and J. Zeitz. *Agricultural Protection in OECD Countries: Its Cost to Less-Developed Countries.* Research Report No. 21. Washington, D.C.: IFPRI, December 1980.

van Lennep, J. E. "North-South Mutual Interest in Greater Reliance on Markets." *World Economy* 5(1982):5–18.

Watkins, M. H. "A Staple Theory of Economic Growth." *Can. J. Econ. & Pol. Sci.* 29(1963):141–158.

World Bank. *World Development Report, 1982.* New York: Oxford University Press, 1982.

Yudelman, M. "Development Issues in the 80's: Achieving Food Security." Paper presented at 26th annual conference of the Australian Agricultural Economics Society, Melbourne, 9–11 February 1982.

Discussion

Alberto Valdés
International Food Policy Research Institute
Washington, D.C.

Robert Thompson has presented us with a comprehensive survey of the ways trade can and should be used to achieve agricultural development and of the barriers to realizing trade potential. He also explores the role trade can play in the achievement of food security goals for less-developed countries (LDCs). He comes through as a strong believer in the real gains from trade-oriented policies for developing countries.

Thompson succeeds in presenting a pedagogical description, in general terms, of the role of trade in agricultural development. I don't disagree with the substance of his presentation. Perhaps my main difference would have been to emphasize more than he did the issues and choices faced by LDCs in the process of reforming their foreign trade regime as it affects agriculture. I believe that although there is substantial agreement with other economists from LDCs at the level at which these general propositions are summarized by Thompson, there is substantial disagreement about the policy prescriptions with respect to trade policy. What explains these differences then? Which major dimensions of the problem are missing from our analysis?

I will first comment on issues related to the trade regime and agricultural development, then follow with brief remarks on food security issues. I believe this is a useful analytical distinction, although of course both are interrelated in a policymaking context. Furthermore, I find that Thompson's analysis of trade issues is the strongest part of his paper; his discussion of food security is interesting, although secondary in the paper.

Trade Regimes and Agriculture in Developing Countries

If we observe the trade regimes of different countries at similar levels of development, we find a variety of trade strategies for food and agricultural exports. This difference is beyond what can be explained by differences in resource endowments and consumption patterns. Faced with the same international markets, the policies of individual countries seem to be consistent with different ways market price signals can be used to develop an economy. These differences persist even though most professional economists in LDCs will agree that long-run trends in international markets are efficient signals for resource allocation, though daily quotations are not. More trade-oriented strategies tend to be followed by countries that have less faith in planners' ability to administer the domestic price system to promote rapid growth. Naturally, international markets then become more important.

We should not underestimate the weight of ideology about the role of markets and governments in the design of agricultural policy. There are those opposed to markets, partly because they dislike the income distribution and the apparent lack of control of economic events as a result of a dependence on market forces. On the other side are those who oppose government (intervention) because of the risk of excessive politicization of economic activities and skepticism about planners' ability to guide economic activities given the enormous complexities of any real life economy.[1]

This discussion is not the appropriate vehicle in which to analyze the role of markets. As I see his paper, it was not Robert Thompson's objective to address the question of why so many developing countries have long been dissatisfied with the international economic system. That they are dissatisfied is unquestionable. And this discontent is directed not only towards international commodity markets and capital and labor markets, but also, and more importantly as each day goes by, it lies in their problems with the international monetary order, especially fluctuating external exchange rates and interest rates. For example, important as it is, trade liberalization is not a panacea for eliminating instability. Monetary disturbances have become very influential (Schuh, Chapter 8). The reality is that, given the growing internationalization of the world economy, welfare in most Third World countries depends increasingly on these international markets, over which they have practically no control, and in a distressingly large number of cases, very little understanding.

Thus, when we look at the role of trade and food security from an LDC perspective, it is difficult to understand and perhaps inappropriate to analyze these issues in the narrow context of agricultural commodity markets. The question of an optimal degree of "openness" of the agricultural economy—of whether LDCs should rely more or less on trade as a source of either food security or agricultural development—

should be seen in this broader context of the international monetary order, services, and commodity markets.

I thought it would be useful here to briefly present a taxonomy of the rational motives for trade intervention, based on simple welfare economics. They present a framework to understanding the arguments for trade intervention. Whether or not these arguments are among the most significant motives for trade intervention in agriculture in LDCs is another question. Unfortunately, I don't believe we have done much historical work to answer this question, but we are not totally ignorant. We have learned something from a few case studies. A simple taxonomy would look like this:

Export Taxes to Help Finance Government

The oldest trade intervention in most LDCs, export taxes continue to be important in countries with a thin tax base, as the collection costs of alternative taxes is disproportionately high. This could be particularly important in the poorest LDCs, and less so with middle-income LDCs. A consequence of these export taxes is a significant squeeze on export activities in agriculture, inducing underinvestment in these activities relative to their potential contribution to overall growth.

Although it is doubtful whether it has much relevance for most LDCs, a different reasoning applies in a few countries, such as Thailand for rice and Bangladesh for jute, namely when export taxes are levied as "optimum" taxes designed to exploit a country's monopoly power. Although not very relevant for most LDCs, the "optimum" tariff argument allegedly applies in the importation of cereals by the EC and Japan, according to Carter and Schmitz (1979).

Trade Restrictions to Reduce Economic Instability

It has been recognized in the debate on export orientation that the degree to which a country chooses to rely on foreign trade may relate to the instability of its export proceeds (McBean 1966, and others). Similarly, short-run instability in world food markets makes these markets appear to be unreliable guides for import planning and long-run domestic production planning. Thus the argument has been made that it would be to the advantage of many LDCs to reduce their reliance on international markets by setting the domestic price of staples somewhat higher than world prices to reduce imports, and hence the vulnerability to world price fluctuations. There is a symmetry between fluctuations in export earnings and fluctuations in the food import bill. The former has been well analyzed in trade theory for a long time, but analysis of the latter has surfaced only recently as part of the debate on food security.

These arguments come up frequently in discussing the role of markets in the world food economy. They apply, for example, in countries like Sudan, where exports are highly concentrated (63 percent from cotton and 23 percent from oilseeds), and thus the country is extremely vulnerable to commodity price fluctuations and adverse world demand trends. On that basis, a case can be made for moving out of those activities with exceptional fluctuations, and diversifying Sudan's exports.[2] Similarly, for food-deficit countries, the argument is that, although it would lower average incomes over the period of a full cycle, it would be to the advantage of many LDCs to set the domestic price of staples somewhat higher than world prices to reduce imports, and hence the vulnerability to world price fluctuations. Although in fact most LDCs do not protect their food production, protection is presented as a desirable goal. Jabara and Thompson's (1980) argument in favor of an optimal tariff under international price uncertainty applies to both situations.[3] Is it a first-best policy though? Probably not. Which are the true constraints then to operate first-best policies, such as fiscal policy of fluctuating reserves? This is an issue that deserves more attention in our research agenda.

Furthermore, although through trade policies domestic prices can be managed in a way that their variability is reduced vis-à-vis the variability in world prices, some empirical evidence, albeit sketchy, suggests that this reduction in real price instability is often not achieved. Two cases come to mind. Empirical tests were done for the comparison between the variability in domestic and world prices for Brazil and Chile. For Brazil, Homen de Melo concludes that during the 1948–1976 period, domestic prices received by farmers for policy-induced home goods (beans, rice, potatoes, and onions) and intermediate products (maize and peanuts) experienced higher relative instability than the corresponding world prices (expressed in domestic currency). Among export crops, instability in the domestic price of coffee was higher than the export price, while domestic price instability in soybeans and cotton was similar to that in world markets. The opposite happened for sugar and wheat, with domestic price instability lower than that for world prices for those products.[4] For Chile during 1961–1974, at a time of administered prices for several products, Varas, Mujica, and Banfi (1977) conclude that domestic price instability in wheat, rice, maize, sugar, beef, and vegetable oil would not have increased if Chile had opened up its economy and let world prices prevail in the domestic market.[5] In the presence of inflation, the task of setting market clearing prices on many commodities and services, and administering varying levels of stocks and imports could become too complex for the administrative capacity of those governments.

Protection and Income Distribution

A principal reason for protection in developed countries has been to prevent changes in income distribution that would have taken place as a result of market forces. An outstanding example of this income maintenance motive is the "senescent industry" argument for protecting continental European agriculture. This argument was made in the late nineteenth century and takes its contemporary form in the common agricultural policy of the European Community.[6] The LDCs' argument for squeezing agriculture through the so-called cheap food policy is the exact opposite of the EC argument. This has been recognized as a method of raising real wages in urban areas at the expense of rural rents, often perceived as a painless extraction from agriculture. The LDCs would squeeze agriculture in order to boost the "sunrise" industries such as steel, petrochemicals, etc. We now know that, unfortunately, the extraction for some countries was rather painful in terms of overall long-run growth in agriculture and the rest of the economy.[7]

Market Access for Exports and Food Imports

As rightly argued by Thompson, restrictions on and uncertainties about market access for LDC exports seem to be major impediments to a more trade-oriented policy in many LDCs. This is usually referred to as export pessimism, and it results in discrimination against agriculture in LDCs. The case of agricultural protectionism in developed countries is well known.[8] Here there is a symmetry between trade liberalization and food security. With respect to security in food supplies, there is the perceived risk of being denied dependable access to foreign supplies as a result of political events, transport bottlenecks in exporting countries, boycotts and dockers' strikes, or unfavorable weather conditions in major exporting or importing countries. Because of this possibility, protective policies to achieve some degree of self-sufficiency in the country's main staples (for which demand is likely to be inelastic) may be justified in terms of security in a strategic commodity. This argument is explicit, for example, in South Korea's rice policy (Anderson 1981). A pertinent question is whether this security could be purchased by means of higher stock levels rather than higher levels of production flows, year in and year out.

Thompson concludes, if I understand him correctly, that the international system would be adequate to provide food security to poor countries if the problems of market access to LDC exports and policy-induced instability in world prices were reduced. Is this all that needs repairing in the international system? I believe LDC concerns about the growing use of economic sanctions by developed countries and world price instability are more acute than perceived by Thompson. The presence of threat from major suppliers is important, even if it has not been effective in the past.

Discouraging Activities Noncompetitive with the World Price

A result of protectionist policies in developed countries is to generate world prices lower than they would be with less protection. Thus, the world price of a commodity tends to approach the domestic price of the exporting country with the strongest comparative advantage. Although the argument cannot logically apply to the whole economy of a country, as the real exchange rate would prevent it, it could apply to individual traded commodities for any country. Protectionism in dairy products and sugar in developed countries are two good examples these days. For both products, a combination of import duties and export subsidies in several developed countries is creating a major downward pressure on current export prices, inducing a movement out of these activities in developing countries. For most of the dairy industry in developing countries, it is hard to compete with imports of dairy products in the absence of protection. On the export side, sugar exporters from LDCs are competing in Third World markets with highly subsidized exports from the European Community. Which is the first-best policy in the long run is not simple, considering that the urban areas would benefit from these cheap imports in dairy products and that protectionist policies could change.

Perhaps a more significant case for low-income developing countries is the one involving food aid. Concessional sales represent an implicit export subsidy that could induce an increasing dependence on imported cereals in the recipient countries, through its potential effects on the structure of incentives to local producers and on lower prices to local urban consumers. Food aid is an unreliable source of supplies in the long run and so recipient countries have a hard time trying to understand what constitutes reliable guidelines for import planning and long-term domestic production policy.[9]

Implicit Taxation on Agriculture

The implicit taxation of exportables and import-competing activities in agriculture results from the indirect effects of a booming nonagricultural export sector and industrial protection.

In oil-exporting countries (this could apply also to mineral exporting countries and to recipients of relatively large influxes of foreign assistance), the external surplus brings currency revaluation or domestic inflation, raising the price of nontradeable goods relative to the price of exports and imports. As a consequence, import-competing (such as cereals) and traditional export sectors are being squeezed.

A similar situation involves industrial protection in LDCs, with its effects on lowering the real exchange rate and its result on disincentives for traded activities in agriculture.[10] These two situations could fall under the self-imposed impediments, as Thompson calls them, which today represent major policy issues in countries with a good resource base for agriculture, such as Nigeria, Mexico, Colombia, and many others. As a disguised devaluation, some countries attempt to defend their agriculture through a combination of import duties and export subsidies on some agricultural products, although tariffs on food products are seldom applied.

I have excluded in this categorization the issue of financing average amounts of food imports, on the grounds that this is not an argument for promotion of import substitution in food, as domestic producers' prices eventually reflect the constraints in the country's import capacity through exchange rate adjustments.

This taxonomy covers, I believe, the most significant arguments submitted by trade theory for trade intervention policies in agricultural commodity markets that concern food security and agricultural development objectives. The infant industry argument has not been included mainly because it is less relevant for agriculture. Are these motives for trade intervention consistent with observed trade policies for agriculture in most LDCs?

First, I don't believe we have done enough historical work to answer this question. Second, although economic science might have a great deal to say about policies, economic goals are not the only goals. Noneconomic considerations are very important, and in many instances they force policymakers to adopt policies they would consider to be far from first-best. As economists, we don't have a framework for dealing with these noneconomic considerations in policy analysis. (Bates and de Janvry's papers—see Part 4—are most relevant here.) Third, in my opinion, there are several gaps in our knowledge as economists that are particularly relevant for the food sector. I would single out the lack of an analytical framework to examine how to bring down the short-run losses resulting from implementation of long run–oriented policies that lead to high efficiency gains. Why is it that so often we don't consider "first-best" solutions to the problem at hand, and instead circle around the "feasible" solutions only—which usually represent the nth best solution anyway? Our knowledge on implementation of policies is very limited. This is, for example, the case with management of short-run instability and its implications for long-run resource allocation in agriculture. Fourth, there is ignorance, and the effective demand for work on agricultural policy is often too low in relation to the importance of food and agriculture in LDCs. On the supply side, I believe the policy analysis capability in several countries is quite weak. Thus, sometimes in LDCs we focus on the wrong issues (such as some of the debate on buffer stocks for LDCs), and sometimes we don't know how to operate effectively in the international system.

An illustration that relates to this last issue is the argument that LDCs are either unable to or underestimate their ability to manage price risks. The argument is that many LDCs are not using some of the institutions available to deal with price risks, such as futures markets, and that this is leading them away from more trade-oriented policies.[11] To what extent does this represent a true "distortion" in economic information, in the sense that many LDC importers are not using available information about world market conditions, which would be worth acquiring after adjusting for the cost of such information? It would be useful to extract the relevant theory and evidence on this point. Personally, I would predict that the results of such a study would indicate a substantial underinvestment in LDCs in institutional arrangements such as market information and the processing of such information. There is an important role for local government in establishing such institutions. Also, international organizations could do a better job, particularly with respect to the low-income developing countries.

Food Security

Thompson states that poverty explains most food insecurity. Thus he stresses a growth strategy to raise per capita income, as a way of reducing food insecurity. We could add that a strong element of food security is a strong world agriculture. I find it impossible to disagree that these are basic elements. However, even when agriculture is growing rapidly, LDCs need to be able to cope effectively with such short-run phenomena as instability in production, world prices, and import capacity, which could separate them from a static view of their optimum long-run trade path. When food consumption is often insufficient even during normal years, supply shortfalls induce a rise in food prices, resulting in major reductions in consumption. Trade in relation to short-term management of food supplies is, I find, not given its share in Thompson's paper.

Thompson's concept of food security seems too broad for practical policy analysis. First I would put aside the issue of acute food shortages caused by natural disasters such as earthquakes and floods, or disasters induced by political upheavals, such as in Cambodia in 1979 and in

Somalia more recently. In those circumstances, the quantities of food required are usually small. These situations require essentially engineering and political solutions, but they do not have significant implications for international markets, and their trade policy implications are analytically straightforward.

In our discussion of food security, I believe we are concerned mainly with how international trade should support the attempts of countries to meet annual targets of food consumption on a year-to-year basis. What constitutes target consumption is, of course, a central issue. In essence, the problem of food security in low-income countries lies in the sharp and unanticipated fluctuations in real income that alter the food consumption of both the rural and urban poor. Trade can help stabilize aggregate supplies, particularly in urban areas where food insecurity is usually associated with fluctuations in food supply. However, in the absence of means to deliver food to rural households exposed to food insecurity—a common situation in LDCs—food supplies may be stable for the nation as a whole but still insecure for large segments of the rural population. So far we have focused on the supply side, yet coping with reductions in rural income due to agricultural production shortfalls represents perhaps the most intractable aspect of food security policies. In solving this problem, trade has little to offer.

Notes

1. For a very useful analysis on the topic, see G. Edward Schuh, "The Role of Markets and Governments in the World Food Economy," Chapter 8 this book.

2. K. Nashashibi, "A Supply Framework for Exchange Reform in Developing Countries: The Experience of Sudan," IMF Staff Paper, Vol. 27, No. 1. (Washington, D.C.: IMF, March 1980).

3. In presenting the case for an optimal tariff or export tax, Jabara and Thompson (1980) are referring to government planners' risk aversion, over and above the pattern of crop diversification chosen by farmers and private decisionmakers in general. Theirs is contrary to the more traditional view, which claims that because the government can spread risk over many activities and sectors, its intervention could help neutralize this risk aversion of producers.

4. F. B. Homen de Melo, "Abertura ao Exterior e Instabilidade de Precos Agricolas," Trabalho para Discussao No. 40, (Sao Paulo: FIPE, December 1980).

5. J. I. Varas, R. Mujica, and S. Banfi, "Politicas de Precios Agropecuarios: Metodologia para Definir una Banda Optima de Precios," Documento de Trabajo 53 (Santiago: Instituto de Economia, Universidad Catolica, 1977).

6. W. M. Corden, *Trade Policy and Economic Welfare* (Oxford: Claredon Press, 1974).

7. A recent study examining these intersectoral linkages is that of D. Cavallo and Y. Mundlak, *Agriculture and Economic Growth in an Open Economy: The Case of Argentina*, Research Report 36 (Washington, D.C.: IFPRI, December 1982).

8. A. Valdés and J. Zietz, *Agricultural Protection in OECD Countries: Its Cost to Less-Developed Countries*, Research Report No. 21 (Washington, D.C.: IFPRI, December 1980).

9. For Egypt between 1956 and 1978, Scobie finds that a 10 percent rise in the world wheat price reduced receipts of concessionary wheat by 4.4 percent. G. M. Scobie, *Government Policy and Food Imports: The Case of Wheat in Egypt*, Research Report No. 29 (Washington, D.C.: IFPRI, December 1981).

10. J. Garcia, *The Effects of Exchange Rates and Commercial Policy on Agricultural Incentives in Colombia: 1953-1978*, Research Report No. 24, (Washington, D.C.: IFPRI, June 1981).

11. For production decisions on crops with long gestation periods, Thompson (see Chapter 7) is explicit when he argues that no country can hedge the prices of traded goods five years in advance when most investments have a payback period of several years.

References

Anderson, Kim. "Northeast Asian Agricultural Protection in Historical and Comparative Perspective: The Case of South Korea," Research Paper No. 83. Canberra: Australia-Japan Research Centre, May 1981.

Carter, Colin, and Andrew Schmitz. "Import Tariffs and Price Formation in the World Wheat Market." *American Journal of Agricultural Economics* 61 (1979): 517-522.

Cavallo, Domingo, and Yair Mudlak. *Agriculture and Economic Growth in an Open Economy: The Case of Argentina*, Research Report No. 36. Washington, D.C.: International Food Policy Research Institute, 1982.

Corden, W. M. *Trade Policy and Economic Welfare*. Oxford: Clarendon Press, 1974.

Garcia, Jorge. *The Effects of Exchange Rates and Commercial Policy on Agricultural Incentives in Colombia: 1953-1978*, Research Report No. 24. Washington, D.C.: International Food Policy Research Institute, 1981.

Homen de Melo, F. B. "Abertura ao Exterior e Instabilidade de Precos Agricolas," Trabalho para Discussao No. 40. Sao Paulo: FIPE, December 1980.

Jabara, Cathy, and Robert Thompson. "Agricultural Comparative Advantage Under International Price Uncertainty." *American Journal of Agricultural Economics* 62 (1980): 188-198.

Knudsen, O., and A. Parnes. *Trade Instability and Economic Development*. Lexington, Mass.: Lexington Books, 1975.

Macbean, A. *Export Instability and Economic Development*. Cambridge, Massachusetts: Harvard University Press, 1966.

Nashashibi, K. "A Supply Framework for Exchange Reform in Developing Countries: The Experience of Sudan." IMF Staff Paper, Vol. 27, No. 1. Washington, D,C.: International Monetary Fund, March 1980.

Scobie, Grant M. *Government Policy and Food Imports: The Case of Wheat in Egypt*, Research Report No. 29. Washington, D.C.: International Food Policy Research Institute, 1981.

Valdés, Alberto, and Joachim Zietz. *Agricultural Protection in OECD Countries: Its Cost to Less-Developed Countries*, Research Report No. 21. Washington, D.C.: International Food Policy Research Institute, 1980.

Varas, J. I., R. Mujica, and S. Banfi. "Politicas de Precios Agropecuarios: Metodologia para Definir una Banda Optima de Precios," Documento de Trabajo 53. Santiago: Instituto de Economia, Universidad Catolica, 1977.

Discussion _____

T. K. Warley
Ontario Argicultural College
University of Guelph

One difficulty I had in preparing to discuss Professor Thompson's contribution was that I found nothing disagreeable in its contents.

Accordingly, I anticipated that I could best approach the issues raised in Thompson's paper by bringing into sharper focus the concrete policies and programs that would contribute to the alleviation of hunger in the world and the enhancement of food security. Although the emphasis is on an inventory of international policy measures, I also list measures that are the responsibility of the national authorities in countries in which hunger exists and food supplies are insecure. This juxtaposition emphasizes that solutions to these problems require a synergistic mix of national and international policy initiatives. It will also be seen that the international policy agenda embraces more than trade measures.

Hunger and Malnutrition

The following measures would find a place on most people's lists of policies to help alleviate the problem of hunger and malnutrition.

National Measures

- Reduce the rate of population increase
- Eliminate absolute poverty
- Increase indigenous food output in developing countries:
  - Accord higher priority to agriculture in development strategies
  - Make larger investments in:
    - Appropriate technology
    - The supply of modern inputs
    - The development of rural infrastructure
    - Human capital

- Remove price disincentives to increase food output
- Lift non-economic constraints on agricultural development
- Raise the capacity to import food:
  - Improve efficiency in the production and merchandising of an appropriate mix of export products
  - Avoid overvalued exchange rates
  - Improve international food procurement practices[1]

International Policy Measures

- Provide more generous and more effective aid:
  - Expand bilateral and multilateral financial and technical assistance to agricultural and rural development programs in developing countries
  - Food aid:
    - Make expanded and longer-term quantitative commitments[2]
    - Tie food aid programs more closely to:
      agricultural and rural development, and
      the feeding of targeted malnourished groups
- Raise LDC food import capacity:
  - Improve the overall environment of the world economy:
    - Restoration of growth with price and currency stability
    - Reaffirmation of the principles and strengthening of the integrity, authority, and capability of the General Agreement on Tariffs and Trade (GATT) and the International Monetary Fund (IMF)
    - Lower interest rates and energy prices[3]
  - Strengthen the LDCs' capacity to cope with balance-of-payments problems:[4]
    - Expand and improve balance-of-payments support mechanisms
    - Permit debt rescheduling for selected countries
  - Raise and stabilize the LDCs' foreign exchange earnings:
    - Abjure protectionism and further open world markets
    - Avoid unfair trade competition with products of export interest to the LDCs
    - Enact, where appropriate, stabilization-oriented international commodity agreements
    - Encourage the use of the IMF's compensatory finance facility
- Reduce the risks to the LDCs of a trade-oriented food supply strategy:
  - In respect of political risks:
    - Avoid economically and diplomatically motivated food export restraints

- Assuage LDC fears about the threat of a grain exporters' cartel
- Avoid the rigidities in world grain markets associated with the proliferation of bilateral trade agreements
- In respect of market risks:
  - Improve the production and delivery capacity of Western agriculture
  - Achieve greater stability in the price and availability of grains in the world market by:
    changing national farm and trade policies so as to allow wider sharing of adjustments in grain consumption and production to changing market conditions; and
    multilateral management of the world grain market

Food Security

This is not a concept that is readily described with a two-word vocabulary. In the long term, enhanced food security requires the assured availability of increasing per capita food supplies. The measures listed above are relevant. The shorter-term concept of food security dealt with here entails developing the capacity to avoid sharp reductions in the food consumption of countries, regions, groups, and individuals due to variation in incomes, food prices, supplies, and availability. Again, both national and international trade and other measures are required.

National Measures

- Stabilize production, especially of poor people's foods
- Sustain incomes of the poor
- Improve national capacities to respond to food emergencies:
  - Develop national early-warning systems
  - Improve food distribution infrastructures
- Operate targeted food subsidy and distribution programs
- Make use of the world food trading system:
  - Maintain financial reserves
  - Improve the handling capacity of ports and internal distribution systems
- Maintain physical food stocks

International Policy Measures

- More effective disaster relief:
  - Improve the operation of the global early warning system
  - Sustain commitments to the international emergency food reserves

- Enhance the contribution of food aid:
  - Maintain the total volumes available in high price periods
  - Introduce a flexible food "insurance" component[5]
- Expand the financial resources of LDCs faced with unusual food situations through:
  - Improved general balance-of-payments support
  - More flexible credit arrangements
  - Further development of the IMF's food import financing facility
- Reduce the risks and instabilities facing LDCs in turning to world grain markets by:
  - Abjuring export restraints by rich grain exporters
  - Wider sharing of reductions in consumption in high price periods
  - Enactment of multilateral arrangements for grain pricing and stocking
- Provide international financial support for a system of developing country–owned security reserves[6]

Some Issues

Among the multitudinous issues raised by Thompson's paper that we could address, I would suggest five that should receive our particular attention.

The central issue is, of course, the appropriate balance between autarkic and trade-dependent food supply strategies. One task is to identify what can be said that is of general and enduring applicability considering the disparate circumstances of developing countries. We should also share views on the future course of the terms of trade between grains and products of export interest to developing countries. The relevant statistic here is not the net barter terms of trade but the double factorial terms of trade.

Secondly, we might explore what can be usefully said about the appropriate mix of the policies identified above that can contribute to the alleviation of hunger and malnutrition and the enhancement of food security. It is apparent that the pursuit of the national policy measures listed is a necessary condition, as it cannot be expected that international measures will suffice of themselves, or even that they will command multilateral support if national authorities are seen to be neglecting their responsibilities to provide their peoples with an adequate and secure food supply. Beyond that, however, which of the candidate international policies are in place and which missing, and how do they rank in terms of cost-effectiveness?

I hold the view that international policy is not animated primarily by considerations of globalism, morality, and human solidarity, but by perceptions of national interest. Accordingly, we need to identify the national benefits that would derive from the implementation of the proposed "international" policy measures.

This is a fortiori true of the role of the United States, for it is still the case that little of consequence can be accomplished in the world without U.S. cooperation or acquiescence. Allowing the United States' central position in the world economy and, more especially, in the world's food economy, what are the particular U.S. interests, responsibilities, and contributions?

Finally, I cannot accept that the role of a Wheat Trade Convention with price stabilization, joint stocking, and shared adjustment provisions—which commanded so much multilateral attention and effort in the 1974–1979 period—should be passed over in analyzing the role of markets in the world food economy. Because world grain markets are seriously flawed, it would be remiss of us not to explore the influence on their functioning (for better or for worse) of an international commodity agreement for grains.

Acknowledgments

I would like to thank Ammar Siamwalla for his helpful comments.

Notes

1. A recent study indicated that developing countries could significantly lower the costs of their grain imports by improving their grain trading practices, in such areas as tender terms, hedging, timing, and payment. See World Food Council, "Cereal Import Procurement Practices and Alternatives for Developing Countries," WFC/1982/5/Add.1 (New York: WFC, March 1982).

2. The figure on page 274 (top) shows the inverse relationship that held during the 1970s between concessional food aid supplies and their opportunity cost to donor countries. The international community pledged at the 1974 World Food Conference to make food aid a more predictable and assured component of poor countries' food supply strategies by making their commitments in quantitative terms. To date, the target of 10 million metric tons annually has not been reached.

3. It has been estimated that every percentage point rise in the LIBOR rate increases the LDCs' annual interest payments on external debt by $1 billion, and each $1 per barrel rise in oil prices raises the annual cost of their oil imports by $2 billion. "Money is Dearer than Oil," *Economist* (20 December 1980), p.65.

4. The magnitude of the debt burden of the non-oil exporting LDCs is indicated in the table on page 274 (bottom).

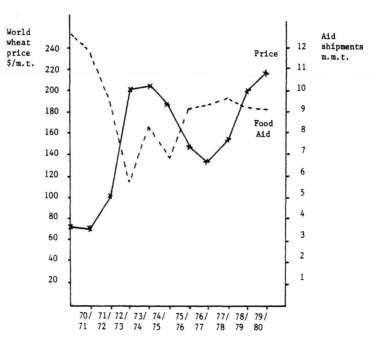

Source: U.S.D.A., Foreign Agricultural Circular, World Grain Situation/Outlook. Various issues.

Non-oil Exporting LDC Debt 1980

	Amount	Average Annual Change 1972-1980
	(Billion $ U.S.)	(Percent)
Debt Outstanding		
Publicly guaranteed	340	21
• Official creditors	158	16
• Private creditors	182	28
Other	74	18
Total	414	21
Debt Service Payments	75	26
Debt Ratios (1979)	Percent	
Debt/GNP	18	
Debt/exports	78	
Debt/reserves	176	
Service as % of exports	13	

Source: World Bank, World Development Report 1982. Washington, D.C.: World Bank, 1982.

As the following tables show, the rising cost of servicing this debt, together with the larger proportion of export earnings required for oil and food imports, is eroding the fraction of export receipts available for the importation of goods and services necessary for economic development, particularly in the net oil-importing LDCs.

Use of Foreign Exchange

(Percent of Export Earnings)

(a) All LDCs			
	1970	1981	1986 (forecast)
Food	8.5	5.5	6.0
Debt service	16.0	17.5	24.0
Oil	3.5	8.5	8.5
Compressible imports	72.0	68.5	61.5

(b) Net Oil-Importing LDCs

Significant Food Importers

	1970	1981	1986 (forecast)
Food	17.0	19.0	20.0
Debt service	7.5	18.5	25.0
Oil	4.5	10.0	9.5
Compressible imports	71.0	52.5	45.5

Source: Adapted from "Future Economic Imperatives," paper by D. E. Smee, Bank of Montreal, at a conference on "The Power of Food," Canadian Institute of International Affairs, Waterloo, Ontario, 6 June 1982.

5. This concept is developed in D. Gale Johnson, "Increased Stability of Grain Supplies in Developing Countries: Optimal Carry-overs and Insurance," *World Dev.* 4 (1976):977–987, and in P. Konandreas, B. Huddleston and V. Ramangkura, *Food Security: An Insurance Approach*, Res. Rep. No. 4 (Washington, D.C.: IFPRI, 1978).

6. This proposal has been made recently by the Secretariat of the World Food Council. See World Food Council, "World Food Security and Market Stability: A Developing Country-Owned Reserve," WFC/1982/5, (New York: WFC, March 1982).

Part 6 ⎯⎯⎯⎯⎯⎯⎯⎯⎯⎯⎯⎯⎯⎯⎯⎯⎯⎯

8

The Role of Markets and Governments in the World Food Economy

G. Edward Schuh

Department of Agricultural and Applied Economics
University of Minnesota, St. Paul

Probably no issue creates more political tension on the international scene than the role of markets and of governments in the world's economy. We live in an ideological world. The respective roles of markets and governments go to the heart of how societies organize their social, economic, and political activities.

Those opposed to markets use various arguments. Some fear markets and the lack of discipline they imply for certain groups in society. Others dislike the income distribution that results from dependence on markets. Still others simply prefer a stronger sense of direction to economic activities than markets imply, especially if the direction of the economy that markets determine is not consistent with their particular values and beliefs. Each of these groups believes governments or the public sector should have a strong role in the allocation of resources and in the distribution of income.

Juxtaposed against these groups are those for whom governments as organizers of economic activities are anathema and who see markets as the only way to organize such activities. For these groups, government intervention leads to a loss of economic efficiency, results in interference in the "natural" forces of markets in distributing income and causes the excessive politicization of economic activities.

The truth of the matter, of course, is that neither of these extreme positions is technically defensible. Clearly, markets can do some things quite well. But there are other things they don't do very well at all. The same is true for governments. Hence, the key question is not whether economic activities will be organized either one way or another. Instead, the important policy issue is to decide what activities should be organized by means of markets and what activities should be organized by the public sector. As we will see later, this is in large part an empirical question, and one that must be resolved in the context of the particular economy and its stage of development.

Not so many years ago, individual countries resolved these questions largely within the confines of their own body politic, through political processes appropriate to that body politic. The world has changed greatly, however, and the issue is no longer such a simple matter. Throughout the post–World War II period, international trade has grown more rapidly than has world GNP. This has led to a growing internationalization of the world's economy. Under these changed conditions, the welfare of individual countries depends increasingly on international trade, which inherently denotes an increased interaction with the international economy.

The original scope for my paper was to be the perspective of individual nations in considering the respective roles of governments and markets. As I worked on the paper, however, I found this perspective confining, and not very relevant in considering the major issues we face in the world food economy. Hence, I broadened my scope to include the more complex task of identifying the role of governments and markets in the international economy.

Other developments in the post–World War II period have been equally important in raising questions about the proper roles of markets and governments. For example, a large, well-integrated international market for capital has evolved over the last two decades that links the various economies of the world together as effectively and as importantly as does trade. Interestingly, this market is almost completely unfettered by government rules and regulations. Equally important, the private banking system seems to have made major errors either in evaluating the repayment ability of those who were borrowing money, or in failing to make any evaluation at all. This has put the entire system at risk.

Another important development was the breakdown of many elements of the Bretton-Woods Conventions, the set of rules and regulations established at the end of World War II to manage the international monetary system. One change was the shift from what was essentially a system of fixed exchange rates to one that is essentially a system of flexible exchange rates. This occurred in 1973, when the United States devalued the dollar for the second time in slightly over two years, closed its gold window and essentially forced the world to a system of floating exchange rates.

Interestingly enough, it was the hope of U.S. authorities at that time that this change would affect another important pillar of the Bretton-Woods Conventions—the dependence on a limited number of reserve currencies. The United States hoped specifically that the dollar would become less important as a reserve currency. In fact, however, the world is still very much on a dollar standard.[1]

Changes in trade patterns have also raised important organizational and institutional questions. For example, the General Agreement on Tariffs and Trade (GATT) was designed and organized largely by the industrialized countries of the West. Moreover, these countries were the main signatories to the GATT. Since trade in the immediate post–World War II period was largely among the industrialized market economies, this created few problems. However, over the last decade the centrally planned and less-developed countries have played an increasing role in international trade. Consequently, a larger and larger share of international trade takes place without the protection and discipline of the GATT.

Finally, attitudes toward trade have changed significantly over time. In the immediate post–World War II period, the industrialized market economies, led by the United States, were very much free trade–oriented. The centrally planned and less-developed countries were both autarkic, wanting to cut themselves off from the international economy and being strongly motivated toward government intervention and control in such trade as they did permit. Today, these positions are almost completely reversed, especially on the importance of freer trade. It is the industrialized countries that are becoming protectionist and interventionist. The centrally planned and less-developed countries, on the other hand, are becoming increasingly outward-oriented and concerned about freer trade—at least with respect to that trade that benefits them.

The less-developed countries have long been dissatisfied with the international economic system that emerged at the end of World War II. Perhaps the first dramatic expression of this dissatisfaction was with the first U.N. Conference on Trade and Development (UNCTAD) back in 1964.[2] Out of a series of these conferences has evolved what is referred to as the North-South debate, a competition between the less-developed countries and the industrialized West, with the primary focus on LDC opposition to the United States.

To date, the "North-South debate" might more properly be called the North-South Dialogue of the Deaf. Neither side appears to pay much heed to what the other is saying, nor to make any semblance of an accommodation that might lead to constructive negotiations. The centerpiece of the South side of the dialogue is a plea for a New International Economic Order (NIEO). This has two main components: (1) a plea for increased income transfers from the industrialized countries, and (2) a demand for international commodity agreements to protect the less-developed countries from what they *perceive* as a chronic tendency for the terms of trade to shift against them.

The North, on the other hand, has not developed a similar cohesive negotiating posture, nor does it appear to have a clear notion of how it might respond to the demands of the South. Until recently, if there was any unifying theme to policy initiatives on the part of the North it was to argue for increased dependence on trade and greater dependence on market forces. As these countries have become increasingly protectionist in recent years, however, even that theme has been muted.

An important premise of my paper is that there *is* a need to reform our international economic order and change how we organize international economic activities. I will describe the major reforms and changes I believe we need, and suggest how we might achieve them. The discussion of these proposed changes and reforms will include a discussion of the principles involved.

Many of the principles are as relevant to domestic economies as they are to the international economy. What I essentially will do is draw on principles that have been developed over time for organizing national economies for the insights they offer to the organization of the international economy.

A couple of caveats are in order. First, although my topic has to do with the world food economy, one cannot discuss the larger organizational issues without considering a broader range of economic activities, such as trade in industrial products. Second, any attempt to change and reform our present international institutions has to be laced with a strong dose of realism. It is one thing to talk about the benefits of free labor and capital markets. It is quite another to move in a discreet fashion from where we now are to where we might be. National entities and national identities will not disappear overnight, nor will we change how economic activities are organized in individual countries. We *can* say something about how countries relate to each other, however. The challenge today is to determine how we might take small steps in the right direction in improving how we relate to each other.

Finally, my discussion will inevitably be sketchy. Neither time nor space permits an analysis and prescription in sufficient depth to lay out detailed guidelines. However, perhaps my paper will motivate some constructive discussion that will put us on the fabled first step of a long march.

The remainder of the paper is organized into four parts. First I will discuss the international monetary order, and second the product markets or trade. The third part concerns the capital and labor markets, and the fourth income redistribution. I close with some concluding comments.

The International Monetary Order

A stable monetary order is generally assumed to be essential for a stable economic order. A stable price level is required if resources are to be allocated efficiently and if capital markets are to perform efficiently. Given the proclivities everyone shares for higher per capita incomes, efficiency is a desirable goal. Given the increased dependence of most countries on international capital markets, the improved performance of these markets is also a desirable goal.

Under the provisions of the Bretton-Woods Conventions questions of the monetary order were largely a domestic or internal issue. Countries were mandated to resolve problems in their external accounts by changing their domestic policies. In this way individual countries were to be precluded from dumping their problems abroad by pursuing beggar-thy-neighbor competitive devaluations, as some of them did during the 1930s.

As long as the supplier of the major reserve currency for the world (the United States) pursued monetary policies that maintained a relatively stable price level, and as long as international capital markets were either atrophied or nonexistent, that system worked reasonably well. But when the United States began to inflate its economy because it was unwilling to raise domestic taxes to finance a war and a massive expansion of its social welfare programs,[3] the game changed. The problem was further complicated by the emergence of international capital markets, and the granting to the International Monetary Fund (IMF) of the right to create international reserves in the form of Special Drawing Rights (SDRs). Much of the world inflation of the 1970s was due to excessive pumping of dollars into the international system by the United States, and the IMF's creation of large amounts of additional international reserves by the mere stroke of a pen. The large increase in the value of gold at this time exacerbated the problem by increasing the monetary value of gold reserves.

Conventional wisdom has it that individual countries can isolate themselves from inflation in the international economy by letting their exchange rates float. Two comments on that "wisdom" appear to be in order. First, when international capital markets were virtually nonexistent, there may have been some basis for that proposition. However, it is seriously weakened by the extent and efficiency of the present international capital market. Second, the consequences of exchange rate realignments are not innocuous. The purchasing power parity doctrine, in which exchange rates reflect only price level differentials, is valid only when international capital markets are unimportant. Once major capital flows occur, exchange rate realignments can induce major shifts in the production sectors of individual countries.

This problem becomes especially serious when the world is subject to major monetary disturbances, as it has been over the last decade or so. Since about 1968, the United States has been an important source of disturbances with its stop-and-go monetary policies. The IMF has also contributed, especially with the significant increase in international monetary reserves in the early 1970s—mostly in the form of newly created SDRs.

The shift to a system of floating exchange rates in the presence of a well-integrated international capital market is of special significance to international commodity markets,[4] particularly in light of the role and importance of the United States in those markets. Under this regime, trade sectors play a much greater role in bearing the consequences of changes in monetary policy. Tight monetary policies in the United States attract capital from abroad, and this bids up the value of the dollar. An increase in the value of the dollar damps off exports and translates international prices into the U.S. economy at a lower level in dollar terms. Similarly, easy money policies cause a capital outflow, which in turn causes the value of the dollar to decline. This stimulates exports and raises agricultural prices in the United States.

Similar adjustments occur on the import side. A rise in the value of the dollar in response to tight monetary policies causes the price of imports to decline in dollar terms. This affects important competing sectors such as the sugar industry, the automobile industry, and steel. Similarly, a decline in the value of the dollar in response to easy monetary policies makes imports more expensive in dollar terms, providing a stimulus to import competing sectors.

Thus one sees that the adjustment of the economy in response to changes in monetary policy significantly affects the export and import competing sectors. Agriculture, as both an important export and import sector, is subject to monetary shocks when monetary policies change. The problem is that U.S. monetary policy has been very unstable since about 1968. An important share of the instability of U.S. agricultural markets in the 1970s is due to this monetary instability. Under the old fixed exchange rate system with an atrophied or nonexistent international capital market these monetary shocks did not occur.

It should be noted that these disturbances in response to unstable monetary policy are not limited to the United States. In the first place, the international exchange rate system can best be characterized as one of block floating. A large number of countries tie their currencies to one of the major reserve currencies, and particularly to the U.S. dollar. In countries that tie their currencies to the dollar, agriculture experiences the same monetary disturbances as does U.S. agriculture. Mexico is an outstanding example of a country that apparently failed to realize the difference between tying its currency to the dollar in fixed and flexible exchange rate systems.

It is also important to note that the United States is a major importer of agricultural products—second only to Japan. Shifts in demand due to monetarily induced realignments of exchange rates impose similar instability on exporting countries. This is true whether the currency is tied to the U.S. dollar or whether it floats.

In a world of perfect resource mobility this change in how monetary policy affects the economy under a flexible exchange rate system might not be all that important. However, resource mobility between agriculture and the rest of the economy is notoriously sluggish. The consequence is an overcommitment of resources to agriculture during some periods, and a serious adjustment problem during others. U.S. agriculture currently illustrates this point quite well. As a consequence of a weak dollar, additional resources were introduced into agriculture during the latter half of the 1970s for the first time in approximately 50 years. Now, with a strong dollar, these resources will most likely have to be transferred back out again. Consequently, agriculture faces a serious adjustment problem.[5] Until this adjustment takes place there will be considerable pressure for protection, for the use of export subsidies, and for price support programs that will have to be protected by the Section 22 waiver. That, of course, is how distortions to free trade become amplified.

More generally, distortions in exchange rates give rise to protectionism on the part of some groups of countries, and to the use of export subsidies on the part of others. They also can create balance-of-payment problems that give rise to demands for income transfers on the part of the less-developed countries, and to plans for market stabilization schemes and other market interventions.

I see little solution to this problem short of establishing an international central bank with a mandate to keep monetary reserves for the international system growing at a measured, steady rate. Such a system could not assure that individual countries would not pursue unstable policies. However, it would remove the onus from the United States of acting like the central bank for the world. It would also reduce the exposure of other countries to the consequences of political pressures on the U.S. Federal Reserve Bank and to the monetary instability which results.

A more stable monetary order is a service that must be provided by the public sector. It is a proper role of government. With a stable order some of the pressures for protectionism will decline and a more favorable environment will be created for reducing barriers to trade. Capital markets will also be able to perform more efficiently, setting the stage for a more efficient allocation of the world's resources.

The Product Markets

Markets are a social invention that provide the means for coordinating the individual efforts and activities of a myriad of individuals pursuing their own individual interests. Hayek[6] makes the important point that we must look at the price system as a mechanism for communicating information if we want to understand its real function. He further argues that the most significant fact about this system is the economy of knowledge with which it operates, or how little the individual participants need to know in order to be able to take the right action.[7] This point is important because in his view the peculiar character of a rational economic order is determined precisely by the fact that the knowledge this system requires never exists in concentrated or integrated form but solely as the dispersed bits of incomplete and frequently contradictory knowledge that the separate individuals possess.[8] This is what Hayek calls the knowledge of the particular circumstances of time and place.[9]

Hayek also notes that there is nothing in this market system that denies the importance of planning. The issue is not whether or not there will be planning. Instead, the issue is *who* will do the planning— whether it will be done centrally, by one authority for the whole economic system, or whether planning is to be divided among many individuals. Competitive markets provide the means for decentralized planning. Whether central planning or decentralized planning will be more efficient depends mainly, in Hayek's view, on which system can make fuller use of existing knowledge.[10] Because of the importance he attaches to knowledge of a particular place and time, Hayek obviously believes competitive markets will be the more efficient.

Economic activities in the United States are organized in large part through markets. There is a large public sector, however, and of course a great deal of economic activities take place in the household. These household activities include an important part of the production of human capital for society. They also include crucial activities such as feeding the nation's population and a great deal of the health care and nurturing that takes place. Moreover, some of the more interesting questions associated with development have to do with the transfer of activities back and forth between the market economy and the household.

Unfortunately, we know very little in a systematic way about the household economy. Casual observation suggests that this component of our economy is quite large. Yet the output of this sector tends not to show up as part of our GNP. And except for a few economists concerned with the formation of human capital and the human capital approach to labor markets, very little analytical work has been directed towards the household economy.

For many observers, the role of the household is peripheral to the issue of the proper roles of markets and governments. I believe it deserves more than casual reference, however, since much of the current political debate in the United States is not over whether economic activities should be undertaken through markets or by government, but whether they should be done by governments or the household. I refer, of course, to the issues of social welfare and education and training. How these issues are ultimately resolved has important implications for the market economy, since some economic activities will exist in the market economy only as long as strong government programs exist. Moreover, scaling down government in the provision of these services will release resources from the government sector either to the market economy or to the household sector.

Despite these caveats, the issue of the role of government in social welfare, schooling, and training programs is very much a choice of whether these activities will be undertaken by government or in the household. The failure to articulate the choices in this way may well lead to bad social policy. Ultimately, it may have a great deal to say about the continued existence of the family unit as we have known it.

To return to my main theme, the role of government in a market economy has been pretty well circumscribed by economists over the years. Aside from the responsibility for establishing a stable monetary order, the main role for government has been seen as providing a stable civil order—policing and justice—and maintaining competitive markets. The main function in the latter case is to break up monopolies and maintain an antitrust posture. An exception to the competitive market rule occurs when economic conditions determine a natural monopoly—cases in which the technical conditions of production are such that economies of size can be realized only with one firm or unit, or when competition could be obtained only with costly duplication. In these cases, desirable social policy involves granting the monopoly and either forcing it to act as a competitive firm by a bounty or tax, or auctioning off the right to the monopoly so as to tax away the scarcity rent.

Another function of government in a market economy is to provide information that helps make the market more competitive. Information helps establish the conditions that economists assume when they show

how markets lead to an efficient allocation of resources. But information is not likely to be produced by the private sector and made readily available to other firms. Given the ease with which information can be passed from one participant in a market to another, it is generally difficult for the producer of such information to recover the costs incurred in producing it. Hence, production and distribution of market information is generally viewed as a proper role of government.

The more general case for government intervention in markets is where there are externalities or clear divergencies between private and social costs or between private and social returns. In the first case, firms or individuals impose costs on the economy or society that they themselves do not incur, as in the case of pollution. In the second case, the benefits to societies of actions taken by firms or individuals is greater than what the individual can reap. That either leads to under-investment in such activities by the private sector, or to their disappearance.

Even when there is a case for government intervention because of divergencies between private and social costs and private and social returns, there remain many questions of precisely how the government should intervene. Taxes and subsidies, for example, can generally lead to an efficient allocation of resources without the government directly becoming involved in the economic activity.

In the case of commodity markets, arguments are often made that governments need to intervene in order to reduce risk and uncertainty. This is often the basis for price-fixing schemes, and for the holding of reserve or buffer stocks by the government.

D. Gale Johnson[11] showed long ago that risk and uncertainty could lead to a less than efficient allocation of resources and thus to a loss in output from a given bundle of resources. The issue again, however, is what the proper role of government should be in such circumstances. The provision of improved information is one way to deal with the problem of uncertainty. This involves more than collecting data and making it available in a timely fashion. Data combined with analysis can lead to valuable information. An important example is outlook information, which attempts to inform producers and consumers of what prices and conditions will be at some future date so that participants in the market can make better decisions today.

The production of this information requires publicly supported analytical groups to process the data and do the analysis. The importance of these groups in the public sector can be readily seen in the case of international commodity markets. Few firms would have the size to gather the data and do the analysis. Even if firms of sufficient size could generate the information, it is not clear that they would be able to recover the costs incurred. This further supports the notion of the public sector providing this service.

Improved information is not the only means of dealing with the risk and uncertainty problem, however. A number of institutions have evolved over the years to provide means of sharing risks and uncertainty or of transferring risks from direct participants in commodity markets to other members of society. One institution is the future market. Future markets do not reduce the instability, risk, and uncertainty in individual markets. They do, however, provide a means whereby producers can stabilize their own price expectations or stabilize their income flow. The role of government vis-à-vis such institutions is to see that they are established, that they work properly, and to guard against fraud.

One of the puzzles on the U.S. scene is that future markets are not more widely used by farmers. Gardner[12] suggests this behavior implies that producers may not be as risk averse as is generally believed. Moreover, farmers appear to want a "reasonable" price guaranteed, with the privilege of gambling for higher prices.

Other examples could be provided. The important point, however, is to note that there is an important role for government in establishing institutions such as those cited above. In general, such institutions will be preferable to direct involvement of the government in economic activities.

Another important case where strong government intervention is often demanded is in the management of reserves or buffer stocks. Pressure for intervention tends to come from three sources: producers in exporter countries, policymakers in less-developed importing countries who dislike the balance-of-payment consequences of unstable prices, and those generally concerned that without adequate reserves there will be famines. This latter is the familiar food scarcity argument.

A number of comments are in order on this issue. First, the argument from producers in exporting countries generally amounts to a plea for price and income *support,* not price and income stabilization. To my knowledge there have been no pleas from such groups for reserves to level out or reduce high prices. The requests come only at times of low prices.

Second, the posture taken by the United States that importing countries should help carry the burden of reserve stocks is misguided. The economics of stock carrying leans to the side of exporters. Importers really have little or no incentive to carry such stocks, and are not likely to do it. For small countries not able to influence world prices by their actions, the rational policy is to carry extra foreign exchange reserves so that they can acquire supplies when they have a domestic shortfall or when world prices are high.[13]

Third, government stocks tend to displace private stock holdings. Consequently, the cost-effectiveness of government stocks tends to be quite low. In addition, the management of such stocks is often destabilizing rather than stabilizing. The problem managers face is knowing when a particular price fluctuation is a temporary aberration and when it is the start of a trend. Because of this difficulty, the managers make mistakes and increased instability is the result.

Finally, commodity stabilization schemes are a costly means to stabilize balance of payments for the less-developed countries. It would be much preferred to rely on the international financial facilities to deal with balance-of-payment difficulties rather than to intervene in international commodity markets.

The issues surrounding stock carrying and food security are important examples of where developments and actions in the international economy lead to externalities that give rise to perceived needs for government intervention. As Gale Johnson[14] has pointed out so effectively, an important source of instability in international commodity markets is the prevalence of barriers to trade and autarchic commodity policies. The failure to let international prices be reflected to domestic producers and consumers precludes the needed adjustment to changing conditions of demand and supply. Consequently, prices respond in an exaggerated fashion to shocks to those markets.

The key to food security and to more stable commodity markets is to reduce the barriers to trade.[15] Given the variety of places in which most commodities are produced, it is seldom that bad weather will affect all of those regions at one time. Hence, in the absence of barriers to trade, weather-induced instability in international commodity markets would be fairly limited. Moreover, there would not likely be a need for buffer stocks other than what normal market forces would induce.

It is important to recognize that trade distortions which cause instability of international markets to be larger than they otherwise would be are of two quite distinct kinds. On the one side of the market are tariffs, undervalued currencies (implicit tariffs),[16] and nontariff barriers to trade. Although not exclusively, such policies tend to be more prevalent among the advanced industrialized countries. It is these countries that tend to protect their agriculture.

On the other side of the market are distortions that shift the domestic terms of trade severely against agriculture. These policies, which tend to be prevalent among the less-developed countries, include overvalued currencies (implicit export taxes), explicit export taxes, export quotas and embargoes, and high levels of protection for the industrial sector. These policies tend to reduce the production capacity in these countries, often causing them to shift from being net exporters to being net importers. They also result in extensive forms of agricultural production, with only limited dependence on purchased imports, or imports produced in the industrial sector. This limited use of modern imports limits the flexibility of agriculture in these countries, and when combined with the reduced production capacity they have, cause more demands on international markets and more shocks to the system.

Reforming the international system and finding ways to deal with these problems have to be high on the agenda for reformers of the current international system. They are collective issues and not likely to be resolved by the initiatives of individual countries. The incentives for action by particular countries are just too limited.

Participants in the most recent round of multilateral trade negotiations had little stomach for the continuation of those negotiations, nor did they believe at the termination of those negotiations that there was much to be gained from another round. But the mutual or collective reduction of trade barriers is the only way sufficient trade-offs can be generated to create the incentives for a reduction in barriers to trade.

The next round of multilateral negotiations should have a significantly broader negotiating agenda if much progress is to be made in lowering barriers to trade. In the first place, the less-developed countries should be brought in as full-fledged participants. This obviously complicates what has already been a complicated set of negotiations in the past. But the role of these countries in trade is now so great they can no longer be ignored. Moreover, their newfound interest in trade makes their participation timely.

Second, the full range of trade distortions should be placed on the table. Export quotas and embargoes are as important as barriers to trade as are tariffs and nontariff barriers. Third, distortions in foreign exchange markets should also be a part of the negotiations. Overvalued currencies may well be the most prevalent distortion to trade. As a tax on exports they have reduced the productive capacity of agriculture worldwide. As import subsidies, they have given rise to the need for high tariff barriers for the industrial sector, especially in the less-developed countries. Very little progress can be made in reducing or eliminating one form of distortion if the other sets of distortions are not addressed.

The problem of undervalued currencies also needs to be addressed. Although used primarily by Japan and to a lesser extent Germany as a general practice, some of the green currencies of the European Community's common agricultural policy also constitute overvalued currencies. This distortion may become increasingly important as countries become more trade-oriented in response to the need to import more petroleum and other scarce raw materials.

An important innovation in the most recent round of trade negotiations was the attention given to the codes of conduct. This emphasis needs to be sustained in any renewed negotiations, with the focus expanded to include the implicit export subsidies that are often reflected in domestic policies.

An important source of pessimism about further trade negotiations is frustration with the GATT machinery for adjusting trade disputes. This machinery has become very bureaucratic and the delays in reaching judgments are long and costly. The solution to this problem is to reform that machinery as well. Change is going to be needed in any case, especially if the number of countries represented is increased and the range of negotiable topics enlarged. We need to get on with the necessary reforms.

Trade negotiations are not the only policy initiatives needed in the trade sector. In the first place, to have trade liberalization, some means need to be found for dealing with trade adjustment problems. Unfortunately, trade problems are all too often articulated domestically as a conflict between foreign and domestic producers instead of being presented accurately as a conflict between domestic consumers and domestic producers. One way to deal with this problem would be by an international approach—for example, through an International Trade Adjustment Fund, financed by a tax or "contribution" based on the value of trade for individual countries. This fund would then be used to deal with adjustment problems created by trade liberalization. It would provide an international means for dealing with what is perceived as essentially an international problem.

I must admit that so far trade adjustment policies have not been very effective. Even the instruments of the 1974 Trade Adjustment Act in the United States have seldom been used in a timely fashion. But our failure to devise such mechanisms so far should not preclude our making the attempt.

Another aspect of the trade situation is the tendency of exporting countries, especially the United States, to use implicit export subsidies in the form of food aid to dump their excess production abroad. Once the dollar was devalued in 1971 and 1973, the need for this subsidy disappeared and food aid declined significantly. In recent years no more than 5 percent of our total exports were shipped on concessional terms.

The dollar has now risen dramatically, and many observers, including this one, believe it is overvalued again as a consequence of the United States playing the role of central banker for the world. The weakening of exports and the resulting accumulation of domestic stocks have generated political pressures to increase our concessional sales again.

The significance of food aid in the context of trade negotiation is that such aid enables governments of less-developed countries to discriminate against their agriculture by means of trade policies. If these policies that shift the domestic terms of trade against agriculture are to be altered, the penalty for pursuing these policies needs to be increased. Continued use of food aid makes it easy for countries to continue distorting their trade policies.

Clearly, there is a role for food aid to assist other countries in times of natural catastrophe, and possibly as a form of developmental assistance. However, as developmental assistance it should be channeled to facilitate investments in human capital,[17] and not to provide a bail-out for balance-of-payment problems or as a means for dealing with recurring production shortfalls at home.

To close this section, it should be noted that trade liberalization per se is not a panacea for eliminating instability from international commodity markets, although it obviously has an important role. The problem of monetary disturbances still needs to be addressed, as it is an important source of instability.

Finally, the discussion in this section has suggested a significant role for what could essentially be called an international government. That role, however, has been limited to setting and monitoring the rules for international trade. If a more effective set of rules were established and enforced, it would give markets a greater role to play on the international scene. Moreover, reductions in barriers to trade on the international scene may well lead to increased dependence on markets and a freer play of market forces within individual countries. A greater dependence on markets would lead to a more efficient use of the world's resources.[18]

The Capital and Labor Markets

Ironically, capital and labor markets seldom are considered in discussions of agricultural policy, yet they may be as important as commodity markets in developing an efficient agriculture. Certainly they are of critical importance in dealing with problems of equity, although in this case their role is again seldom recognized.

In taking our bearing on labor and capital markets it is important to consider the changes that generally take place as agriculture is modernized and as an economy develops. Perhaps the predominant feature of this process—what Bruce Johnston[19] has called the one universal rule of economic development, is that labor has to be transferred out of agriculture. Parallel to that transfer is the need for an increase in the use of capital in agriculture. New technology is imbedded in new imports, more modern imports are used, and the stock of capital per worker has to increase if per capita incomes of rural workers are to rise. Consequently, if agriculture is to be modernized, and if farm people and workers are to earn incomes comparable to those earned in the nonfarm sector, labor and capital markets (including that for land) have to perform efficiently.

Unfortunately, governments tend to intervene extensively in the wrong way in capital and labor markets, and fail to intervene in the proper way. As a basis for discussion it is useful to consider appropriate ways for governments to intervene in factor markets. Given the need to facilitate mobility and prevent large income disparities between the rural and urban sectors, it is generally recognized that providing market information is an important role for the public sector. Although word of mouth is an important source of information in labor markets, there is still an important role to be played by employment services that help employers identify where unemployed workers are located and to help workers identify where alternative employment exists.

Under certain circumstances a case can be made for subsidizing labor mobility. Given that labor moving out of agriculture to alternative employment typically has to move geographically as well, the costs— both psychic and pecuniary—can be significant. Affecting or reducing these costs can make for a more efficient allocation of resources, an important externality, and thus can be justified as a proper role of government. Certainly it is more desirable than intervening in commodity markets as a means of offsetting the consequences of a low rate of migration.

Another proper form of government intervention in labor markets is the investment in formal schooling and training of the labor force. Capital markets to produce human capital tend not to be efficient, in part because there are important externalities associated with education. Yet schooling, for example, has been found to be an important means of accelerating migration from agriculture.[20]

It is important to note that an important cause of the apparent premature migration from agriculture is the tendency of governments to shift the domestic terms of trade against agriculture by large distortions in trade policy.[21] Reducing such distortions is an important means of reducing the outmigration and in turn the clogging of intersectoral labor markets that has been so characteristic of many less-developed countries.

In addition, it should be recognized that due to the high proportion of young, well-educated, and entrepreneurial migrants, there may well be important negative consequences for the region supplying the migrants. When combined with the problems associated with large concentrations of people in urban agglomerations, there may be a case for government intervention to decentralize the industrialization process.[22] These interventions will reduce the need for geographic mobility while increasing intersectoral mobility. This can make for a more efficient allocation of the nation's resources, and also for a more equitable distribution of income.

The proper role of government in credit markets is to create the capital market instruments necessary to encourage savings at appropriate levels—a much neglected aspect of policy in most countries—and to provide the institutional arrangements that permit loan funds to be extended to producers at the social cost of those funds. An important issue on this side of the market is that transaction costs are often large for loans extended to small producers. Consequently, bankers and other issuers of credit tend to neglect this sector. As an alternative, highly subsidized funds are often provided to this sector. A more desirable policy may be to find ways of offsetting the transaction costs directly rather than to provide the subsidized credit.

A proper role of government in the land market is to provide proper cadastres of the land and to help assure that titles are secure. In countries where inheritance laws have led to excessive fragmentation, there is also an important role to be played in consolidating land holdings. Care should be exercised, however, to not interfere in fragmentation that serves to spread risk by holding multiple parcels located in different geographic areas with different production potential.

The importance of maintaining an open land market is a seriously neglected goal of policy in many countries. In fact, government intervention often impedes the transfer of land ownership and promotes fragmentation. Such policies fail to recognize that an increase in farm size is an important means to increase the per capita income of rural people, and an inherent part of the development process. As per capita incomes rise in the nonfarm sector, farms need to become larger if incomes in the farm sector are to increase at the same pace. Within limits, inputs other than land can be added to labor to raise its productivity. At some point, however, economic forces will dictate that additional land is needed, leading to farm enlargement.[23]

Distortions in labor and capital markets are an important source of the duality that emerges in labor markets in many less-developed countries, and in the open and hidden unemployment that emerges in these economies. These factors are not insignificant in U.S. and other industrialized economies, however.

It is not uncommon in less-developed countries to find subsidized credit and subsidized imports of capital goods used as the primary means of promoting economic development, and minimum wages and high payroll taxes used as the means to deal with perceived equity problems, especially in urban labor markets. In the latter case, the payroll taxes are used to support social welfare programs, on the mistaken premise that it is the capitalists who pay these taxes. The truth of the matter is that the incidence of such taxes tend to fall on the worker in the form of unemployment.

In any case, the consequence of highly subsidized credit and a combination of minimum wage laws and high payroll taxes[24] is to shift relative factor prices to induce a highly capital-intensive development trajectory. The distorted factor prices also induce the use of production technologies that are not appropriate for the local resource endowment. Although the tendency is to attribute antilabor or conspirational motives to the capitalists that use such technology, it is really government policy that is to blame. An important by-product of such policies is also a highly skewed or unequal distribution of income. Again, the tendency is to blame the workings of a market economy as causing unequal distribution of income, when in fact government policies are to blame.

The issue, of course, is not whether to subsidize industrialization or not. The issue is *how* to subsidize it. Subsidized education and training programs, for example, can be an important subsidy to private industrialists and farmers. It also can be an important means of dealing with the equity problem, as we will see below.

Government interventions in land markets are as severe as they are in the other two factor markets. Limits are set to farm size, share tenancy is precluded by law, share proportions are determined by government decree, and limitations on land transfers are imposed. All of this interference with market forces impedes the efficient use of resources. More often than not, the resulting distribution of income is diametrically counter to the intent of the policies.

Translating these policy prescriptions to the international arena involves more than a few complexities. Foreign capital is viewed with more than a little suspicion in most countries. Barriers to migration among countries are generally quite severe. And laws preventing the ownership of land by foreigners are quite common.

There are a number of encouraging signs on the international scene, however. For example, a very efficient international market for capital has emerged over the last two decades. Although governments are reluctant to let foreign firms make direct investments in their economy, they have been more than willing to go into international capital markets for credit. This means of financing development programs has largely supplemented concessional foreign aid by other countries as a source of capital funds. This system has been put at risk by the failure of banks and other lending institutions to look after their own best interests, but government intervention is not the solution to this problem. More astute lending is.

International labor markets have also become more open, sometimes by force majeure and sometimes as a rational response to market forces. The petroleum-rich Middle Eastern countries with their sparse population are important examples of the latter. The Mexican border and the boat people of Southeast Asia are important examples of the former.

Properly specified welfare functions that assess the gains a country experiences from economic intercourse with the rest of the world include the factor markets.[25] The exchange of capital and labor can be as important a source of national welfare as exchange of goods and services.

Barriers to the international migration of labor and to international ownership of land are likely to continue into the near future. Perhaps the best that can be done is to keep the market for capital open and efficient, together with a freer flow of trade. The combination of these two factors can lead to a more efficient use of the world's resources. It can also lead to a more equitable distribution of income.

The changing age pyramid for many of the industrialized countries may put substantial pressure on governments to liberalize their labor markets. The same applies to the centrally planned economies. We may well see more international migration of labor in a few years than we ever thought possible a few years ago.

Redistributing Income

Redistributing income is a proper role for governments. In fact, some have argued for a division of labor in which markets are used to allocate resources while governments make whatever marginal changes in the distribution of income are desired by the body politic. Although this is a somewhat simplistic view, it does have at least two technical bases. First, there is no ethical justification for the distribution of income that results from the operation of a market economy. The distribution will depend importantly on the initial distribution of assets, including those of human capital, and that is in large part a chance phenomenon.

Second, there is no technical basis for saying that one distribution of income is better than another. Our inability to make interpersonal comparison of utility means that we really cannot say whether one distribution of income is better than another except in relation to a political goal determined by a political process. Hence, changes in the distribution of income need to be brought about by a political process—by government intervention.

Questions *can* be raised about the means that governments use to bring about changes in the distribution of income. Some of these questions can be raised on the basis of casual empiricism—from observing how past techniques have performed in redistributing income. More recently, attempts have been made to work out more formal criteria for determining whether the means used to redistribute income have been efficient.[26]

An important point about policies designed to redistribute income is that they often have consequences counter to the intended effects. Examples are legion. Minimum wage legislation is a common example; although designed to make workers better off as a class, it often creates unemployment and lower incomes for large numbers of workers. The growing evidence from land reforms and land redistribution schemes suggests that they often do not help the landless worker they were designed to benefit.[27] And price support programs have been found to benefit those who are already relatively well off in the agricultural sector, without benefiting the disadvantaged—presumably the objective of the price policy in the first place.[28]

All too often governments intervene in market forces as a means of redistributing income. One problem in doing this is that it keeps markets from doing what they do best—allocating resources in an efficient manner—while failing to obtain the desired income distribution goal. An important reason for this disparity between intent and result is that the ultimate beneficiaries of the policies tend to be quite different from what superficial expectations suggest.

Transparency is a desirable goal of economic policy. The problem with many government interventions in markets is that they provide implicit subsidies and impose implicit taxes. In general, private firms and private individuals like to receive their subsidies in explicit form, but governments like to impose implicit taxes. The political process drives the system towards the kind of market interventions and attempts to redistribute income that cause a rather inefficient distribution of income, leading to wasteful uses of a nation's resources.

One way to avoid these difficulties is to make transfers explicit and outside the market place. This will become increasingly important as the world's economy becomes increasingly internationalized and we become more concerned about the subsidies governments provide through domestic policies.

Another difficulty in dealing with income distribution problems is the general failure to recognize the amount of resources required for attaining a more equitable distribution of income. In general, income distributions tend to be skewed, with a relatively small number receiving large incomes and a relatively large number receiving lower incomes. Consequently, one can take all the wealth or income away from the well-to-do and still have only a nominal impact on the income of the disadvantaged. When one takes account of the disincentive effects of such redistributive schemes, their desirability as appropriate schemes declines substantially.

An important dimension of this problem occurs in low-income countries where the problem is mass poverty, not a matter of a small group of disadvantaged unable to compete in competitive markets. Moreover, the problems of mass poverty usually involve generalized low productivity. Reducing this poverty entails finding means to raise productivity, not redistribute income.

This raises another important issue. There is a popular view that a more equitable distribution of income can be obtained only at the expense of a loss in resource efficiency, and that a reduction in average per capita incomes is an appropriate price to pay for obtaining a more equitable distribution of income.

This perspective can be used to justify government intervention in market forces. However, internationally or domestically, this approach is rather mischievous. The implied trade-off between equity and efficiency is in general a false dichotomy. Policy instruments are available to improve the lot of the poor without distorting resource use and without sacrificing growth in average per capita incomes.

The key to reducing poverty is to raise the productivity of the disadvantaged. In many low-income countries, this typically will require the diffusion of new production technology[29] on a generalized basis within society. It will also involve investments in the various forms of human capital—nutrition, health, formal schooling, training programs, etc. In general, the social rates of return to such investments are quite high. Consequently, they can lead to a reduction in poverty without sacrificing the growth in per capita income. In fact, they may well increase aggregate growth rates while at the same time reducing poverty.

Finally, it should be noted that Marxist doctrine, with its emphasis on the class struggle, has caused the problem of poverty to be cast rather infelicitously in the context of a *relative* income distribution problem. The problem of absolute poverty therefore tends to be neglected, as does the progress that often occurs in improving the absolute income of the poor.

An important example is Brazil, where there has been a distortion of the policy and political debate on the income distribution problem. During the period of rapid growth in per capita incomes associated with the economic "miracle" of the late 1960s and early 1970s, there is some evidence that the distribution of income became more unequal.[30] This change in the relative distribution of income became the focus of internal political debate. Seldom was it recognized that the absolute income of the poor had improved very substantially during this period—in fact, as much as in any country in the world.[31] Gary Fields contrasted the Brazilian experience with the Indian experience because India has had a more equal distribution of income high on its policy agenda. He found that during the period in which the distribution of income in India was becoming more equal, the absolute income of the lower income classes actually declined. One can leave it to the poor to decide which of these two situations they would prefer.

Professor Schultz and Rati Ram[32] have called our attention to a rather neglected aspect of the income distribution problem—the enormous increase in life expectancy that has occurred in the less-developed countries since World War II. This increase in life expectancy has a number of important implications. First, because it tends to be concentrated among the poor, it is prima facie evidence that the income of these groups has tended to improve. Second, it is a neglected aspect of relative income distribution that in a very real sense reflects a more equal distribution of income. Third, an increase in life expectancy is an important inducement to increased investment in human capital. Hence, it lays the groundwork for further increases in per capita income for these groups.

In terms of the theme of this paper, there obviously is a significant role for government in adjusting the incomes of the population it represents and the relative distribution of that income. Desirable interventions focus on those measures designed to increase the interest in human capital, and in assuring that access to such investments is widely distributed. There is a generally recognized disparity between the private and social rates of return on such investments. Moreover, capital markets often work less efficiently for the disadvantaged than they do for the advantaged. Hence, it is important to focus publicly supported human capital programs on the disadvantaged.

Having said that, it should be recognized that human capital programs are not a panacea for the income distribution problem. In many countries, it is the upper-income groups that capture the subsidies for human capital. When they do, it can well lead to a more unequal distribution of income. Similarly, we understand only poorly the income distribution consequences of economic development. We do have evidence that development induces a more human capital–intensive configuration for the economy.[33] Whether the technology that results will value particular forms of human capital more highly is at this point an open question. Within the range of development experience, however, the evidence is that public investments in a wide range of human capital can help produce a more equal distribution of income, especially if there is broad access to such public investments in the economy.

What does this analysis imply for the international economy? The implications are quite important, especially in terms of the particular form that foreign assistance and other concessional transfers of capital among countries should take. Foreign assistance, or concessional transfers of capital more generally, might well be limited to or concentrated on investments in human capital. In many countries, the underinvestment in human capital is severe. Moreover, providing support for human capital programs need not imply foreign intervention in the educational systems of other countries.

More attention has been given in recent years to increasing the international capacity for agricultural research, both bilaterally and multilaterally. These efforts should be expanded and resource transfers for physical capital reduced. The externalities from human capital are high. All countries, including the donor countries, can benefit from such investments. A reduction in the productivity differentials among countries is the key to reducing the income differentials. It may well be that more equal investments in human capital are the only effective means for narrowing the gap among countries in a reasonable time.

Concluding Comments

The growing internationalization of the world's economy makes it imperative that we reform our international institutions so that we can conduct our business in a more efficient and business-like fashion. Criticizing the arguments of those who want to change the system and rejecting their requests out of hand will not suffice. We need to engage them in a dialogue and to work towards more serious negotiations.

The controversy over the proper role of markets and the proper role of government will continue into the future. However, we need to press for stronger and more effective international government and for more open markets.

Notes

1. R. I. McKinnon, "Currency Substitution and Instability in the World Dollar Market," *Am. Econ. Rev.* 72, 3 (June 1982): 320–333.

2. For an analysis of this early initiative, see Harry G. Johnson, *Economic Policies Toward Less Developed Countries* (Washington D.C.: Brookings Institution, and London: George Allen and Unwin, 1967).

3. An important advantage the issuer of the world's reserve currently has is that it can collect a tax from the world's economy by pumping the system up with its money.

4. For a more detailed discussion, see G. Edward Schuh, "Floating Exchange Rates, International Interdependence, and Agricultural Policy," paper presented at the Meetings of the International Association of Agricultural Economics, Banff, Alberta, Canada, 3–12 September 1979.

5. For more detail, see G. Edward Schuh, "U.S. Agriculture in Transition," testimony before the Joint Economic Committee of the U.S. Congress, April 1982.

6. Friedrich A. Hayek, *Individualism and Economic Order.*

7. Ibid., p. 86.

8. Ibid., p. 77.

9. Ibid., p. 80.

10. Ibid., p. 79.

11. D. Gale Johnson, *Forward Prices for Agriculture* (Chicago: University of Chicago Press, 1947).

12. Bruce Gardner, *The Governing of Agriculture,* Lawrence, Kansas: Regents Press, 1981.

13. Julio A. Penna, "Optional Storage and Export Levels of a Tradeable Product and Their Relationship with Annual Price Variability: The Case of Corn in Brazil," Ph.D. dissertation, Purdue University, Lafayette, Ind., 1974.

14. D. Gale Johnson, "World Agriculture, Commodity Policy, and Price Variability," *Am. J. Agri. Econ.* 57,5 (December 1975):823–828.

15. These barriers cause the monetary disturbances discussed earlier to have an exaggerated impact on commodity markets.

16. Japan has persistently undervalued its currency since the early 1960s; Germany has also, but to a lesser extent. Certain of the Green currencies of the EC have also been undervalued.

17. For suggestions along this line, see G. Edward Schuh, "Food Aid and Human Capital Formation," in *Food Aid and Development* (New York: Agricultural Development Council, 1981).

18. An important but neglected issue in this section is how to incorporate the centrally planned economies into trade negotiations and how to make them more effectively a part of the world economy. These issues deserve a rather extensive paper in their own right.

19. B. F. Johnston, "Agriculture and Structural Transformation in Developing Countries: A Survey of Research," *J. Econ. Lit.* 8 (June 1970):369–404.

20. Micha Gisser, "Schooling and the Farm Problem," *Econometrica* 33(July 1965): 582–592.

21. See Mauro de Regende Lopes, "The Mobilization of Resources from Agriculture: A Policy Analysis for Brazil," Ph.D. dissertation, Purdue University, Lafayette, Ind., 1977.

22. See Michael Lipton, "Migration from Rural Areas of Poor Countries: The Impact on Rural Productivity and Income Distribution"; and G. Edward Schuh, "Out-Migration, Rural Productivity, and the Distribution of Income," in *Migration and the Labor Market in Developing Countries,* ed. Richard H. Sabot (Boulder, Colo.: Westview Press, 1982).

23. Peterson and Kislev have found that most of the increase in farm size in the United States can be explained by such a response to the increase in the price of labor in the nonfarm sector. See Willis Peterson and Yoav Kislev, "Prices, Technology, and Farm Size," *Journal of Political Economy* 90, 3 (June 1982):578–595.

24. In the case of Brazil, for example, such distortions have been huge. A combination of usury laws and high rates of inflation have caused negative real rates of interest as high as 50-70-90 percent. When combined with a severely overvalued currency—an import subsidy for capital goods—the incentive to use a capital-intensive production process is quite great. Payroll taxes, on the other hand, have at times been as high as the supply price of labor, thus badly distorting the price of labor as well. See Morris Whitaker, "Labor Absorption in Brazil: An Analysis of the Industrial Sector." Unpublished Ph.D. dissertation, Purdue University, Lafayette, Ind., 1970.

25. Antonio Brandao, "The Terms of Trade and the Welfare Gains from Trade: New Perspectives," Ph.D. dissertation, Purdue University, Lafayette, Ind., 1978.

26. Gary Becker, "A Theory of Political Behavior," University of Chicago, CSES Working Paper 006-1, September 1981.

27. For data on the Chilean case, see Alberto Valdés, "The Transition to Socialism: Observations on the Chilean Agrarian Reform," in *Employment in Developing Nations,* ed. Edgar O. Edwards (New York: Columbia University Press, 1974), pp. 405–418.

28. See Bruce Gardner, *The Governing of Agriculture* (Lawrence: Regents Press of Kansas, 1981).

29. Professor Schultz has made the case for new production income as a source of income streams. See *Transforming Traditional Agriculture* (New York: Columbia University Press, 1964).

30. I say "some evidence" because very little attention was given to the quality of the data on which the analyses were based.

31. Gary Fields, "Who Benefits from Economic Development? A Reexamination of Brazilian Growth in the 1960s," *American Economic Review* 67, 4 (September 1977): 570–582.

32. Rati Ram, and T. W. Schultz, "Some Economic Implications of Increase in Life Span with Special Reference to India," *Economic Development and Cultural Change* 27, 3 (April 1979): 399–421.

33. G. Edward Schuh, "Economics and International Relations: A Conceptual Framework," *Am. J. Agri. Econ.* 63, 5 (December 1981): 767–778.

Discussion

Emery N. Castle
Resources for the Future
Washington, D.C.

Dr. Schuh has given us one of his wide-ranging, comprehensive papers on the world food economy. He has demonstrated his grasp of international food relations as well as impressive knowledge of intracountry agricultural policies. I always benefit from reading what Schuh has written and this paper is no exception. Therefore, my remarks will supplement rather than criticize. To this end I discuss four issues:

1. The role of markets in addressing scarcity.
2. The declining real cost of foodstuffs and economic growth.
3. The complementarity of government and markets as contrasted to the competitive nature of the two.
4. The case for pragmatism in addressing markets and government.

I believe the four are closely related, and my remarks will make their relationships clear.

The Role of Markets in Addressing Scarcity

It is my belief that markets are a powerful means of addressing scarcity. For 30 years my organization, Resources for the Future, has studied questions of scarcity in the area of natural resources and environmental quality. For those natural resource commodities that are priced in the marketplace, the real cost over time has fallen consistently.

Yet for many natural resource commodities that are not priced in the marketplace, such as certain attributes of the natural environment (air and water quality and atmospheric quality) and certain kinds of outdoor recreation, the real cost may have been rising. Public sector activity can and has increased the supply, in many instances, but not necessarily at declining real costs. Of course, natural resources such as Old Faithful and Crater Lake are unique, and neither public nor private enterprise will duplicate them nor expand their supply. In such cases, the problem is one of providing for such unique natural resource services over time, and providing for close but not necessarily perfect substitutes.

A comparable lesson is learned from the study of food and markets. Those countries that have utilized the market have a far better record in combating hunger than have those that have rejected it. This leads to the issue of the real cost of foodstuffs over time.

The Declining Real Cost of Foodstuffs and Economic Growth

The paper by D. Gale Johnson (chapter 1) provides evidence that the real price of key food commodities has been declining over time, not rising. Furthermore, it is the market-oriented countries that are making these commodities available in world trade. Yet, many analysts argue that the real cost of agricultural products in the future is likely to rise, not fall, a reversal of past trends. Some are also assuming a continuation of global economic growth. Although it is possible both will occur simultaneously, it is not clear that those who predict both have thought through the relation of one to the other. In much of the developing world, a large part of the resources are used for food production. The source of much past economic growth has been a more efficient agriculture that has permitted resources to be released to the rest of the economy. If an increase in the real cost of food products should occur simultaneously with economic growth on a global basis, it will mean an increasing gap in income levels among the lower-, middle-, and upper-income countries. However, I doubt such a development will occur and question if real income will rise on a global basis unless there is also a substantial increase in agricultural productivity. To accommodate economic growth, the real cost of foodstuffs in the developing countries will need to fall or, at least, will not increase sharply. An improved agriculture often is not *sufficient* for economic growth, but either an improved agriculture or access to an efficient agriculture is a *necessary* condition.

Should there be rising real costs for food and declining or stagnant global economic growth, the problem will be much larger than just improving the diet of the malnourished or undernourished. The problem will grow as their numbers increase.

The Complementary Role of Government and Markets

If markets are a powerful means of overcoming scarcity and lowering real costs of foodstuffs, it is logical to ask how governments might enhance market performance. Much of the literature—including the papers presented above—has tended to emphasize the antagonistic nature of the two. There is little doubt that governments can destroy market incentives; there should also be little doubt that governments can enhance market performance.

The history of U.S. agriculture is instructive in this respect. Publicly supported research has stimulated technical change, and industries have developed in response to this new knowledge to supply farmers with new technology and to market their output. Purchased inputs have increased and the farmer's share of the consumer dollar has fallen. Inputs to farmers were made cheaper and more abundant because of publicly supported research, education, extension, credit, and land programs. Government programs also stabilized output prices. Agribusiness flourished as publicly supported research yielded many new products for it to produce, as it utilized personnel trained in publicly supported institutions, and as extension personnel helped it sell its products.

I believe the United States has been well served by this mix of activity, but if a nation embarks on such a course, it should also be committed to economic growth and should be willing to address the resulting income distribution effects. The deeper meaning is that there is an infinite range of government-market mix possibilities. It is far from an either/or proposition. There has been considerable discussion of the detrimental effects of government actions that distort agricultural product and factor markets, and such issues obviously are important. But government can also improve market performance through research, reliable information programs, creation and protection of particular property rights, and education.

The Case for Pragmatism

Schuh begins his paper with a plea for pragmatism and proceeds accordingly. A few words to provide a rationale for this position would appear to be in order.

In 1957–1958 Francis Bator wrote two highly influential articles on markets. One, entitled "The Simple Analysis of Welfare Maximization," appeared in the March 1957 issue of the *American Economic Review*. The other, "The Anatomy of Market Failure," was published in the *Quarterly Journal of Economics* in August 1958. These articles provided an intellectual basis for the identification of conditions under which markets would yield less than optimum results based on a theoretical ideal. Writing some 20 years later in the *Journal of Law and Economics*, Charles Wolf published an article entitled "A Theory of Non Market

Failure: Framework for Implementation Analysis" (April 1979). Wolf also used Pareto optimality as a standard for nonmarket performance and, similar to Bator, deduced conditions that would cause nonmarket activities to "fail."

Significant developments had occurred during the two decades that elapsed between the two articles. An enormous development in quantitative techniques permitted economists and other analysts to manipulate many more variables than had previously been possible. This led to optimistic expectations as to the ability of economists to specify the conditions of optimum performance with either private or public activity.

The other development was a widespread dissatisfaction with the performance of government. By the time Wolf's article was written it was clear that governmental action would not necessarily be superior to markets, even though a market violated some of the technical conditions for optimum performance. It also became clear that it could be very misleading to judge either public or private performance against a theoretical ideal. When judged in this way, it was possible that both would "fail." What, then, is one to do? The correct answer is the one most pragmatic people would give on the basis of common sense and good judgment. The performance of one should be compared with the performance of the other rather than against a theoretical ideal such as "Pareto Optimality."

When the issue is framed in this way we are encouraged to consider the full range of government activities, not just, say, the creation of a government bureau to produce mail deliveries in opposition to letting the market produce firms to compete for mail deliveries. There are a wide range of policies that can and are used to establish the conditions under which markets operate. If viewed in this way, markets and governments are not considered as discrete alternatives but rather substitutes that can be combined in numerous combinations over a very wide range.

This is why the resource allocation effects of markets often should not be separated from their income distribution effects. Although I agree with Schuh as to economists' inability to say one distribution of income is superior to another on normative grounds, I believe economists often can say something about the effect of different distributions of incomes on the subsequent performance of markets. In particular, economists can determine whether one distribution of income is more likely than another to bring about economic stagnation rather than economic growth. When this is done it tells us a great deal about the effect of different kinds and varying degrees of government intervention.

Some of those engaged in the political process may be motivated primarily by philosphical considerations that lead them to favor either the market or an intrusive government on doctrinaire grounds, but surely there are many others who view the problem quite differently. To them the struggle is over economic assets and income streams over time. The kind and extent of government activity and the role given the market are some of the outcomes of the struggle. Schuh cites examples of how such struggles become counterproductive.

The pragmatic philosophy stated by Schuh at the outset of his paper rests on the assumption that economic theory does not provide criteria that will yield ready answers as to some optimum or ideal mix of government and market activity. Rather, the two often should be viewed as mutually reinforcing rather than competitive. Furthermore, there are many different possible combinations of public and private activity that may be considered. Because there are few in the political fray who strive to promote the national as opposed to their special interest, policies often need to be evaluated both from a national as well as from a particular point of view over time. The analyst or the economist is not likely to be able to argue that one alternative is superior to all others. But he or she may be able to provide information and suggest institutional arrangements so that those political participants, even when acting to promote their short-run special interest, do not select counterproductive long-run policies. Even if they should wish to do so, they should be stopped by others who have access to the same information. The main contribution of Schuh's paper is that he identifies a set of national and international policies that would avoid these mistakes.

Discussion

Richard J. Sauer
Minnesota Agricultural Experiment Station
University of Minnesota, St. Paul

I greatly appreciate the opportunity to react and respond to Dr. G. Edward Schuh's paper. At first I wondered, "why me?" given the topics to be discussed and the backgrounds and expertise of those presenting papers. After all, in my training and experience as a biologist and entomologist with a focus on integrated pest management, I developed a strong orientation towards the production aspects of agriculture. Thus, I at one time thought that most of our agricultural problems could be solved through increased and more efficient production. I no longer have that narrow and biased perspective.

An important consideration that Dr. Schuh should have included in his paper, but did not except by implication, is that agricultural research is a very important public sector activity, which has had and will continue to have a significant impact on the world food economy. For example, Dr. Schuh pointed out that a function of the public sector (government) in a market economy is to provide information that helps make the market more competitive. In fact, the provision of improved information is one way to deal with the problem of uncertainty. This involves more than collecting data and making it available in a timely fashion. The data must be followed by analysis to provide useful information. The role of public sector research is critical to improving our methods of analysis. In addition, we need to give increased attention to better systems for delivering this analyzed data to farmers and other clientele. Recent developments in computer-based delivery systems by state extension services will play a major role here. Universities also have a role in educating those who will provide these services in USDA and other federal agencies, in extension, and in the private sector.

Given our current commodity surpluses and corresponding low prices, in my role of securing adequate state, federal, and private resources to maintain the quality and continuity of our research program I frequently encounter critics who argue that they need no further gains in production. Instead they want help in marketing what they already produce for an adequate price. When crop commodities are in surplus, emphasis shifts from technologies that might increase production to tactics and policies that might increase the price per unit of the commodity. This is true for farmers, legislators, and perhaps even research administrators. The extent to which current surpluses prevail and affect support of production research can greatly affect the accuracy of any predictions I might make about production-oriented research achievements between now and the year 2000.

The rate of gain in production has declined in recent years, and I believe it is only reasonable to expect this because the level of production per unit area of land is already so high. The closer we approach the apparent upper limits of a biological system such as crop production, the smaller the gains will be and the greater the cost of attaining these gains will become both in dollars and manpower.

In considering where research may take upper midwestern and U.S. agriculture by the year 2000, I think we must recognize the climate-dependent nature of our production. Although we are blessed with some of the most productive soils in the world, production from these soils is very much dependent on precipitation. Thus, we must allow for the possibility that climatic conditions in at least parts of this country (to say nothing of elsewhere in the world), will be less favorable for agricultural production than those that have prevailed during our lifetime.

We should also recognize the possibility that for at least several years into the future, and perhaps into the year 2000, agricultural production might remain stable or even decline. The cost of production inputs must be reduced or the price of farm products must increase, or many farmers will go out of business, particularly those with recent capital investments financed at high interest rates. One could argue that the only real control the farmer has in this equation is the cost of production inputs, and the only means of controlling this is to reduce the amount of inputs. The extent to which the farmer must reduce purchases of production inputs will influence whether production remains stable or declines.

In spite of serious constraints to increased agricultural production between now and the year 2000, I believe there can be increases in the productive potential of crops and livestock. Whether that production potential is attained will depend upon continued adequate support of related public sector research. Whether that production potential is translated into production on the farm will depend on economics.

I am not only concerned with how the research we are *doing* will affect U.S. agriculture by the year 2000, but also with our failure to address certain issues critical to our economic, social, and political organization. More specifically, in response to and in agreement with Schuh's paper, we have seen a growing internationalization of the world's economy. Any extrapolation into the future suggests more of the same, perhaps at an accelerated rate. This means that unduly concentrating on the production side of agriculture, at least within Minnesota and the Midwest—the area with which I am most familiar—will have little influence on the welfare of farmers in the years ahead. The agricultural economy of Minnesota and the Midwest, and of the entire United States in some aspects, is facing and will apparently continue to face enormous shocks and adjustments from economic, political, and technological forces operating outside national confines. To deal with that changing world we need to understand it. To date I must admit that we are doing relatively little in our research program to generate the knowledge needed to adjust our economy and our resources to this rapidly changing world.

Dr. Schuh has pointed out that U.S. agriculture is in bad shape largely because the value of the dollar has risen so dramatically. That rise is due to our own economic policies, but also due to policies and technological changes in other countries. No amount of new production technology will by itself help U.S. farmers escape their present plight. In fact, in the short term it may make things worse. Yet, I am not ready to concede that we should cut back on or discontinue our current production-oriented research, though considerable redirection in some areas is highly desirable.

At the same time, I recognize that changes in policies, both here and elsewhere, can help a great deal. Yet we continue to underinvest in policy research. We underinvest in attempting to understand developments in agriculture in other countries, and we underinvest in understanding the economic, political, and social forces that are changing both our own society and the larger world society of which we are a part.

Let me shift briefly to another challenge. If we review the progress of U.S. agriculture over the past 100 years, we can credit the foresight of nineteenth century leaders and legislators for establishing what has come to be known today as the land grant university system, with its three-way mission of teaching, research, and extension. I wish to focus particularly on the decentralized state agricultural experiment station network and its counterparts, the state extension services that have served as the primary vehicle for transmitting the results of research to farmers. The result is that U.S. farmers have become outstanding producers of quality food with a firm scientific and technological undergirding. However, just as we can take the credit for helping our farmers become such efficient producers of food, we can be blamed for the lack of success in marketing and distributing that food in a way that ensures a reasonable rate of return for the farmer.

Assuming we can begin to do the necessary marketing and public policy research to remedy this glaring deficiency, how will the output of that research be delivered to the appropriate clientele—government agencies, legislators, agribusiness organizations, and others—so that it will actually serve as an objective basis for changing state, federal, and especially international policies? The traditional extension service system does not seem to be equipped for the task without radical change. I pose this as a question and a challenge, without having the answer.

I offer one closing comment in response to Schuh's paper, much of which makes sense to me. Just as farmers appear to want a reasonable price guaranteed with the privilege of gambling for higher prices, so has Dr. Schuh proposed that we need to press for stronger and more effective international government while at the same time pressing for more open markets. Both sound like a "have your cake and eat it too" philosophy, and one is tempted to slough this off with the reaction that it is not possible. Perhaps we need to recognize that it may be possible, and begin doing the research that we hope will result in the dual goals being not only possible but indeed probable and achievable.

Abbreviations

APC	Agricultural Prices Commission
BMR	basal metabolic rate
CAP	common agricultural policy
CIF	Cost, Insurance and Freight
CIMMYT	Centro Internacional de Mejoramiento de Maiz y Trigo (Mexico) [International Maize and Wheat Improvement Center]
DRI	Integrated Rural Development
EC/EEC	European [Economic] Community
ERS	Economic Research Service
ESCS	Economics, Statistics, and Cooperatives Services
FAO	Food and Agriculture Organization (U.N.)
FDA	Food and Drug Administration
FTC	Federal Trade Commission
GAO	General Accounting Office
GATT	General Agreement on Tariffs and Trade
GDP	gross domestic product
GNP	gross national product
GOL	grain-oilseed-livestock
GPO	Government Printing Office
HANES	Health and Nutrition Examination Survey
HYV	high-yielding variety
IBRD	International Bank of Reconstruction and Development (World Bank)
IFPRI	International Food Policy Research Institute
ILO	International Labor Office
IMF	International Monetary Fund
LDC	less-developed country
LIBOR	London Interbank Offer Rates
NIC	newly industrializing country
NIEO	New International Economic Order
NSS	National Sample Survey (India)

OECD	Organization for Economic Cooperation and Development
OPEC	Organization of Petroleum Exporting Countries
PAN	Program of Nutritional Assistance
PCM	protein-calorie malnutrition
P.L.	Public Law
SDR	Special Drawing Right
UNCTAD	U.N. Conference on Trade and Development
USAID	U.S. Agency for International Development
USDA	U.S. Department of Agriculture
WFC	World Food Council
WHO	World Health Organization

Index